ALLE · ZEIT · WACH · 1842

Progress in
Sensory Physiology 3

Editors:
H. Autrum D. Ottoson E. R. Perl R. F. Schmidt

Editor-in-Chief: D. Ottoson

W. D. Willis:
Control of Nociceptive Transmission in the Spinal Cord

With 51 Figures

Springer-Verlag
Berlin Heidelberg New York 1982

Editor-in-Chief:

Professor Dr. David Ottoson
Karolinska Institutet, Fysiologiska Institutionen II
Solnavägen 1, S-10401 Stockholm 60

Editors:

Professor Dr. Hansjochem Autrum
Zoologisches Institut der Universität München
Luisenstraße 14, D-8000 München 2

Professor Dr. Edward R. Perl
Department of Physiology
University of North Carolina at Chapel Hill
Chapel Hill, NC 27514 (USA)

Professor Dr. Robert F. Schmidt
Physiologisches Institut der Universität
Röntgenring 9, D-8700 Würzburg

Author:

Professor Dr. William D. Willis
The Marine Biomedical Institute
University of Texas Medical Branch
Galveston, TX 77550 (USA)

ISBN-13: 978-3-642-68568-2 e-ISBN-13: 978-3-642-68566-8
DOI: 10.1007/978-3-642-68566-8

Library of Congress Cataloging in Publication Data. Main entry under title: Progress in sensory physiology. Includes bibliographies and index. Contents: 1. [without special title – – 3. Control of nociceptive transmission in the spinal cord/W. D. Willis. 1. Vision-Physiological aspects. 2. Senses and sensation. I. Autrum, Hansjochem. II. Ottoson, David, 1918–. [DNLM: 1. Sensation – Physiology – Periodical. 2. Neurophysiology – Periodical. WI PR681G] QP475.P89 612'.8 81-4430 (U.S.: v.1) AACR2

Printing: Beltz, Offsetdruck, Hemsbach/Bergstraße
Bookbinding: J. Schäffer OHG, Grünstadt

2121/3140-543210

Contents

1 Introduction: Centrifugal Control of Sensory Pathways

1.1 General Background

Sensory experience depends upon neural processing that has both passive and active components (Hagbarth 1960; Gibson 1966). The passive components involve the interaction between a stimulus and sensory receptor organs, the transduction of sensory information into patterns of nerve impulses in afferent nerve fibers, and the transmission of coded sensory information along central neural pathways to interpretive centers in the brain. The active components include the modulation of transmission in sensory pathways by centrifugal control systems originating within the brain and spinal cord, as well as the exploration or avoidance of environmental stimuli by motor-sensory behavior.

The possibility that there might be centrifugal systems for controlling sensory transmission was recognized by several investigators early in this century (e.g., Head and Holmes 1911; Brouwer 1933). Anatomical connections from the brain to the retina and to the olfactory bulb were described by Cajal (1909) and those to the cochlea by Rasmussen (1946), establishing a morphological basis for efferent control of the special sense organs. The discovery of the fusimotor system by Leksell (1945) demonstrated that somatic sensory receptors, such as the muscle spindle, could also be controlled by the central nervous system (CNS). During the 1950s, a series of investigations, both anatomical and physiological, provided a firm experimental basis for the concept of centrifugal control systems in the vertebrate nervous system (Galambos 1956; Hagbarth and Kerr 1954; Hernández-Peón et al. 1956a; Hernández-Peón et al. 1956b; Kuypers 1958; Lindblom and Ottoson 1953, 1956; Walberg 1957). Parallel work on invertebrates demonstrated not only that comparable systems are in operation in these animal forms, but that the simpler nervous systems of invertebrates offer useful preparations for the analysis of the mechanisms of centrifugal control (Alexandrowicz 1951; Kravitz et al. 1963; Kuffler and Eyzaguirre 1955). These mechanisms have been found to include not only postsynaptic excitation and inhibition, but also presynaptic inhibition (Burke and Rudomin 1977; Dudel and Kuffler 1961; Eccles 1964; Schmidt 1971). During the past several decades, work on the centrifugal control systems has progressed rapidly. Much of this effort has been concerned with the special senses or with the control of muscle spindles; however, there have also been a number of studies on the control of central somatosensory pathways (see reviews by Dawson 1958; Hagbarth 1960; Hernández-Peón 1955; Livingston 1959; Schmidt 1973; Towe 1973; Willis and Coggeshall 1978). The present discussion will focus on the centrifugal control systems concerned with one aspect of somatosensory transmission, nociception. Various facets of this subject have been reviewed by others (Basbaum 1981; Basbaum and Fields 1978; Beaumont and Hughes 1979; Besson et al. 1981; Bowsher 1976; Cannon and Liebeskind 1979; Fields and Basbaum 1978; Kerr and Wilson 1978; Long and Hagfors 1975; Mayer and Price 1976; Melzack 1973; Messing and Lytle 1977; Sherman and Liebeskind 1980; Terenius 1978; Teschemacher 1978; Yaksh and Rudy 1978).

1.2 Functional Implications of Centrifugal Control Systems

Centrifugal controls can operate at any level along a sensory pathway. In some cases, efferent pathways reach the sensory receptor level, while in other instances the first control point is the presynaptic terminal of the primary afferent fiber or the second-order neuron of the sensory pathway (Fig. 1; Schmidt 1973). Centrifugal controls at these early levels of sensory pathways should be especially effective in filtering out unwanted information. However, controls are also exerted at higher levels of sensory pathways, including the thalamus.

A traditional way of describing the operation of a sensory pathway is by the stimulus-response relationship. For relatively uncomplicated sensory systems, this relationship can often be expressed as a power function (Stevens 1970; Werner and Mountcastle 1965) or fitted by some other curve, such as a $\log_{\mathrm{tanh}}$ function (Knibestöl 1975; Naka and Rushton 1966). In the case of nonlinear sensory systems, it may be useful to do a more complex analysis, for example, using Wiener kernels, of the responses to quasi-random stimuli (Marmarelis and Naka 1973).

Whatever method is used to describe the stimulus-response function, it is then possible to use a similar analysis to demonstrate changes induced by the action of centrifugal control systems. In the case of simple, linear systems, some of the changes that have been shown for the somatosensory system include a parallel shift in the stimulus-response curve and an alteration in the slope of the curve (Fig. 2; Carstens et al. 1980a). Another possibility is a change in the configuration of the curve. These changes resemble ways in which feedback loops can alter the operation of electronic circuits. For example, a parallel shift in the stimulus-response function implies a change in set point or threshold without a change in gain. On the other hand, a change in the slope of the stimulus-response function is analogous to a change in the gain. An alteration in the shape of the stimulus-response curve may resemble the action of a filter. For example, the transmission of information from low-threshold sensory receptors may be favored (Oliveras et al. 1974a) or, conversely, that from high-threshold receptors can be preferred (Coulter et al. 1974), depending on changes in the response properties of neurons in a sensory pathway.

The behavioral significance of the centrifugal control of sensory pathways has been discussed in a number of contexts. It has long been recognized that passive movements of the eyes are associated with a sense of eye movement, whereas during active movement of the eyes there is not such a sensation. It has been suggested that this difference is the result of a neural discharge produced during active eye movements that inhibits sensory discharges resulting from the eye movements. This discharge is sometimes called a "corollary discharge" (Sperry 1950), and it is thought to cancel unnecessary inputs that could be predicted to

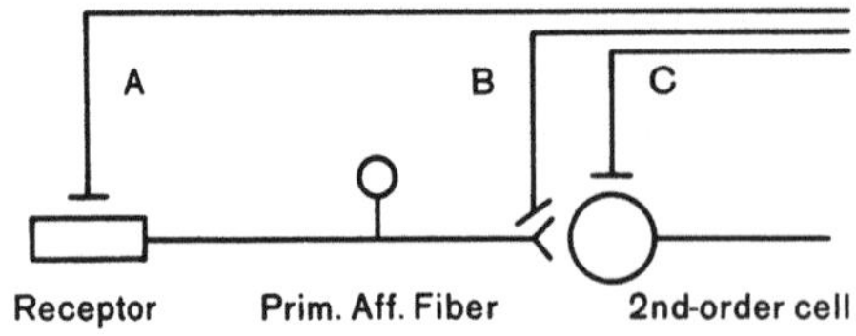

Fig. 1. Possible sites of action of centrifugal control systems originating in the spinal cord or in the brain on initial part of a sensory pathway. (Schmidt 1973)

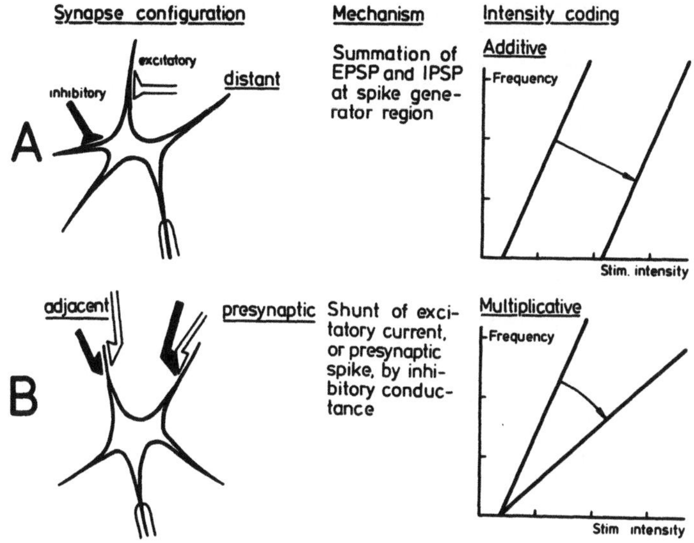

Fig. 2A, B. Two ways in which a stimulus-response curve can be altered by a centrifugal control system. In **A**, there is a parallel shift in the curve, indicating a change in threshold. A mechanism for this might be summation of EPSPs and IPSPs generated at different sites on the neuron. In **B**, the slope of the curve is reduced, indicating a change in gain. Mechanisms might be shunting of an EPSP by an IPSP or presynaptic inhibition. (Carstens et al. 1980a)

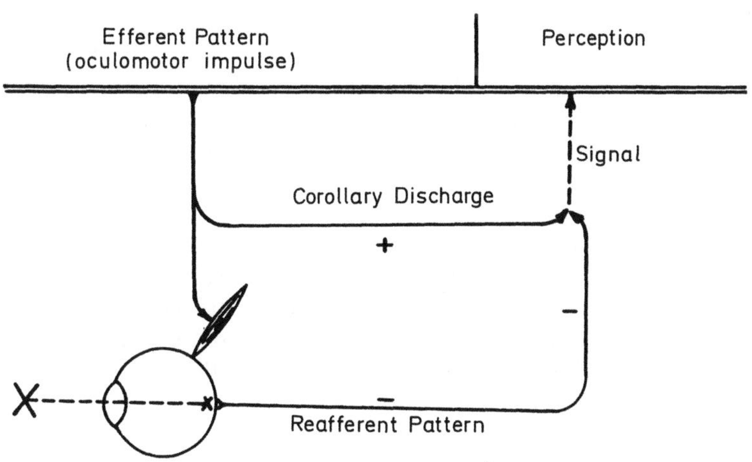

Fig. 3. Diagram illustrating the role of a corollary discharge in the control of eye position. The efferent command is directed not only to the eye muscles but also to a summing point in the central nervous system, where it cancels the input resulting from the eye movement. Thus no signal results to affect perception. (Teuber 1960)

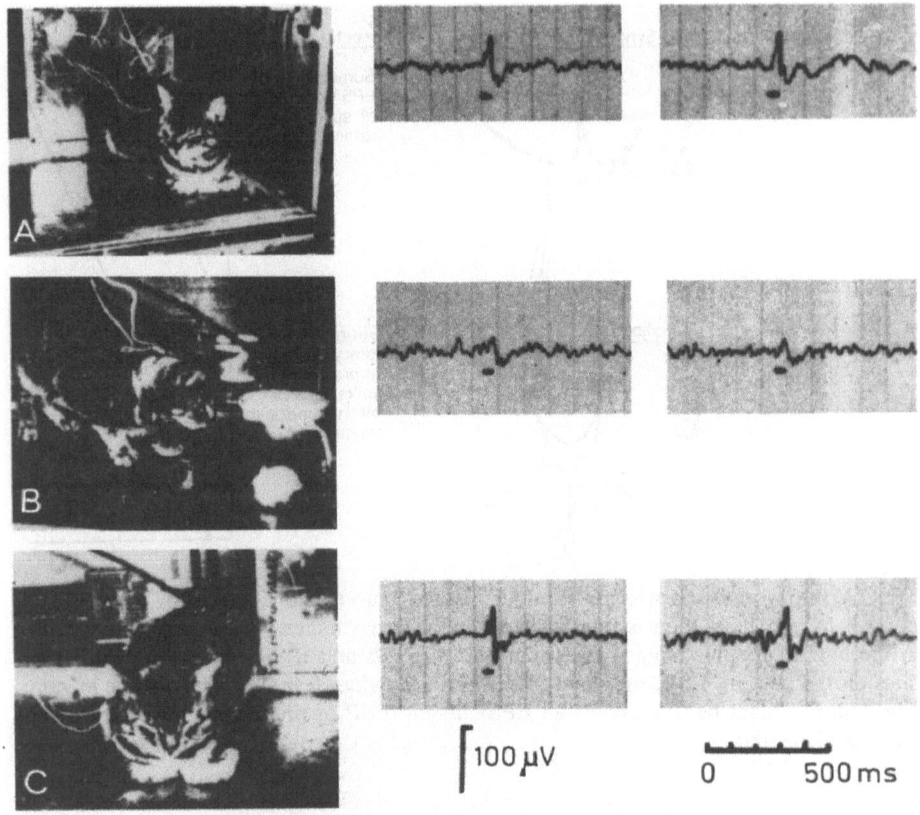

Fig. 4A – C. Changes in evoked potential in the cochlear nucleus during attention. In **A**, the cat is not attending to anything in particular, and the evoked potentials are large. In **B**, a mouse captures the attention of the cat, and the evoked potentials are decreased. In **C**, the cat and the evoked potentials resume their control state. (Hernández-Peón et al. 1956)

result from the movement commanded (Fig. 3; Evarts 1971; Holst 1954; Teuber 1960). Another function of centrifugal control may simply be the elimination of unwanted sensory input. Much sensory data can be regarded as noise, and so inhibiting this input, often at the first synapse by presynaptic inhibition, would serve to enhance the signal-to-noise ratio of potentially important messages (Levitt et al. 1964; Schmidt 1973). Furthermore, it would be important to prevent inputs that might otherwise interfere with motor programs (Dyhre-Poulsen 1975). Although it is still unclear what the true value of sleep is, filtering sensory input might be of advantage during sleep (Carli et al. 1967a, b; Favale et al. 1965). Other functions that have been suggested are roles in habituation and attention (Fig. 4; Hernández-Peón et al. 1956a, 1957; reviewed by Hagbarth 1960; Hernández-Peón 1955; Livingston 1959).

1.3 Somatosensory Pathways Known to be Subject to Descending Controls

It is likely that all the somatosensory pathways are under the control of centrifugal pathways. The best-studied pathway in this regard is the dorsal column-medial lemniscus system. The neurons of the dorsal column nuclei receive synaptic connections from the sensorimotor cerebral cortex (Chambers and Liu 1957; Gordon and Miller 1969; Kuypers 1958; Kuypers and Tuerk 1964; Levitt et al. 1964; Walberg 1957; Weisberg and Rustioni 1976, 1977) and from the reticular formation (Sotgiu and Margnelli 1976; Sotgiu and Marini 1977). Stimulation of the sensorimotor cortex may either inhibit or excite neurons in the dorsal column nuclei (Andersen et al. 1964b; Gordon and Jukes 1964; Jabbur and Towe 1961; Levitt et al. 1964; Towe and Jabbur 1961; Winter 1965). In addition to postsynaptic inhibition, such stimulation produces primary afferent depolarization (and presumably presynaptic inhibition) of the terminals of the axons ascending in the dorsal columns (Andersen et al. 1964a). Primary afferent depolarization is thought to be mediated by axoaxonal synapses (Walberg 1965). Stimulation in the area of the nucleus gigantocellularis of the reticular formation causes an inhibition of neurons of the dorsal column nuclei (Cesa-Bianchi et al. 1968; Cesa-Bianchi and Sotgiu 1969). Behavioral studies show that transmission through the dorsal column nuclei is inhibited during the performance of motor acts (Coulter 1974; Ghez and Lenzi 1971; Ghez and Pisa 1972), suggesting the possibility that a corollary discharge is operating to modify sensory information ascending in this pathway during voluntary activity (Evarts 1971; Holst 1954; Teuber 1960).

In addition to the dorsal column–medial lemniscus system, other somatosensory pathways are also under descending control. For instance, the spinocervical tract is under the influence of several descending control systems (Brown 1971; Brown et al. 1973, 1977; Fetz 1968; Hong et al. 1979; Taub 1964). Both the dorsal column–medial lemniscus path and the spinocervicothalamic system are thought to be involved primarily in mechanoreception, although a role in nociception is also possible (see review by Willis and Coggeshall 1978). Thus the descending control of these pathways is likely to affect the sensations of touch-pressure and flutter-vibration, but there may also be an alteration in pain sensation, at least in animals like the cat (Kennard 1954).

A somatosensory pathway that is thought to be important in nociception, especially in primates and in humans, is the spinothalamic tract (Foerster and Gagel 1932; Vierck and Luck 1979; White and Sweet 1955; Yoss 1953). Experimental studies show that the spinothalamic tract in both cats and monkeys is controlled by descending pathways originating in the brain stem or cerebral cortex (Beall et al. 1976; Coulter et al. 1974; Gerhart et al. 1981a; Giesler et al. 1981a; Haber et al. 1978, 1980; McCreery and Bloedel 1975; McCreery et al. 1979b; Willis et al. 1977; Yezierski et al. 1982). Presumably, descending modulation of the spinothalamic tract would result in significant alterations in the responses to painful stimulation.

The sense of pain is like other sensory modalities with respect to the basic properties of the neural pathways transmitting nociceptive information. An important quantitative difference between pain and other somatosensory modalities is its usually great variability, a fact that is emphasized by Melzack in his book *The Puzzle of Pain* (1973). A particularly dramatic example of the lability of pain sensation was revealed in the study by Beecher (1959) of soldiers wounded in battle during World War II. There were remarkably few complaints of pain due to wounds on the battlefield, perhaps because in the context of war the wounds meant a face-saving escape from the front. The same wounded men complained on inept veni punctures in a normal fashion. It is likely that this variability in pain sensation is attributable in part to the action of descending control systems that modulate transmission in nociceptive pathways (although variability in sensory processing at higher levels of the nervous system is also probable).

1.4 Plan for the Review

Because of the clinical importance of pain, there is considerable contemporary interest in the implications of the centrifugal control systems for pain relief. This review is an attempt to summarize what is known about this topic at present and to indicate some of the directions of current research in this extremely active field of investigation. Section 2 will be concerned with a documentation of the behavioral evidence that there are neural pathways originating in the brain that control nociceptive transmission at the level of the spinal cord. First, there is a brief overview of some of the procedures used to test for analgesia in animals and man. It will be noted that the main tests that relate to spinal cord mechanisms are the flexion reflex and recordings from individual nociceptive neurons. Several forms of "analgesia" in animal models, as well as in humans, can be attributed to descending pathways. However, other types of analgesia may depend more on supraspinal events than on descending pathways. The pharmacology of analgesia due to the action of descending pathways will be discussed in Section 3. Descending control of the flexion reflex is the topic of Section 4, and descending control of nociceptive spinal cord neurons the topic of Section 5. The last section, section 6, is a discussion of correlations between analgesia studies and the descending control of the flexion reflex and of nociceptive neurons.

1.5 Conclusions

1. Sensation depends upon passive and active processes. Among the active processes are centrifugal modulatory actions.
2. Centrifugal control systems operate on the pathways for special sensations but also on somatosensory pathways. Comparable systems are found in the vertebrates and in the invertebrates.
3. One somatosensory system subject to centrifugal control is that responsible for nociception.
4. Centrifugal controls often operate near the input side of a sensory path, either directly on the sensory receptor organ, on the primary afferent terminals, or on the second-order neuron. However, there are also controls at higher levels.
5. Centrifugal actions can change the gain or the threshold of a sensory path or function as a filter.
6. Centrifugal control systems may have important behavioral roles in such phenomena as corollary discharges, enhancement of the signal-to-noise ratio, prevention of interference with motor programs, sleep mechanisms, habituation, and attention.

7. Somatosensory pathways known to be subject to centrifugal controls include the dorsal column–medial lemniscus system, the spinocervical tract, and nociceptive pathways, such as the spinothalamic tract.
8. Centrifugal control of nociceptive transmission is likely to be particularly important both as an explanation of the variability of reactions to noxious stimuli, and as a means for improving therapy for pain relief.

2 Behavioral Evidence for Descending Control of Nociceptive Transmission

2.1 Measures of Analgesia

2.1.1 Background

A major problem in the study of pain mechanisms is the development of suitable ways to measure pain. This problem is difficult enough in human subjects, who are able to describe their sensory experiences, but it becomes even more difficult in animal subjects. Animals are unable to describe their experiences, only to react to them. Furthermore, it is often necessary to use anesthetized preparations to investigate certain nociceptive mechanisms. Anesthesia by definition means that the experimental subject experiences no sensations, including pain. Investigators using animal subjects may thus be confronted with the challenge that they are attempting to study pain in the absence of pain. On the other hand, ethical considerations make it necessary to design experiments on unanesthetized animals with great care. A particularly problematic area is work on models for chronic pain. This topic is extremely important from the standpoint of human disease, and yet there are serious questions concerning animal welfare on the one hand and human welfare on the other. Efforts have been made to develop appropriate guidelines for such studies (Bowd 1980; Covino et al. 1980; Hoff 1980; Iggo 1979; Sternbach 1976; Wall 1975; 1976). The discussion here will be concerned with tests of nociceptive responses. Emphasis will be placed on tests that are appropriate for the study of descending control of nociceptive transmission in the spinal cord, since this is the theme of this review.

A standard term that will be used throughout this review because of its frequent application in the literature is "analgesia". The definition of analgesia is the absence of pain (Greek, painlessness). There are at least three reservations about the widespread use of the term. In the first place, it is common to refer to changes in the degree of analgesia, when levels of hypalgesia are meant (Collier 1962). Analgesia, like pregnancy, is all or none. The second difficulty is that no one can say whether an animal is in pain or not. A way around this terminological problem is to use "antinociception" instead of "analgesia," but this is an awkward term. The third problem with the use of the word "analgesia" is that some of the most common tests for a reduced response to noxious stimulation involve flexion reflexes. To call a reduced reflex response "analgesia" can be regarded as semantic liberty. For example, the tail flick test is a very useful "analgesia" test, yet a response can be produced in an animal following transection of the spinal cord (Irwin et al. 1951). Clearly, an increase in the tail flick latency by, say, morphine in a rat with a spinal cord transection is not a sign of analgesia. On the other hand, it has been found in the intact human that the intensity of perceived pain correlates well with the strength of the flexor withdrawal reflex (Hardy 1953; Price 1972; Willer 1977; however, cf. McGrath et al. 1981; Starkenstein 1932; Willer et al. 1979). Furthermore, analgesic drugs, especially the opiate drugs, often have similar suppressive actions on reflex responses in animals and pain in man (Grumbach 1966). It is for these reasons that the flexion reflex is often regarded as a test of analgesia in animals with an intact neuraxis.

2.1.2 Reflex Tests

Some of the most commonly used tests of analgesia in animals involve various manifestations of the flexor withdrawal reflex (Dewey et al. 1969; Grumbach 1966; Irwin et al. 1951; Martin W. R. et al. 1964). Specific examples of this generalization include simple withdrawal from pinches (pinch test), the tail flick test, the back skin twitch test, the jaw-opening reflex, and the writhing response test. There is a discussion of the neural circuits underlying flexor withdrawal reflexes in Section 4.

A number of investigators have used the pinch test to demonstrate the development of analgesia (e.g., Akaike et al. 1978; Basbaum et al. 1977; Collier 1962; Jacquet and Lajtha 1976; Liebeskind et al. 1973; Mayer and Liebeskind 1974; Mayer et al. 1971; Oliveras et al. 1974a, 1977b, 1979; Yeung et al. 1977). The skin over an extremity is pinched and the response noted. The intensity of the pinch can be quantitated to a certain extent by applying a known force to a known surface area. However, it is not easy to quantitate the output. An advantage of this test is that it can be used to survey the extent of analgesia over the body surface.

A particularly useful test in small rodents is the tail flick test (D'Amour and Smith 1941). This has been used frequently in analgesia studies (e.g., Akil and Liebeskind 1975; Bonnycastle 1962; Dostrovsky and Deakin 1977; Grumbach 1966; Hayashi and Takemori 1971; Hayes et al. 1977a; Irwin et al. 1951; Mayer and Liebeskind 1974; Proudfit and Anderson 1975; Yaksh et al. 1976b; Yeung et al. 1977). The tail is stimulated with radiant heat. The percentage of animals that respond with a flick of the tail within a defined period of time, say 20 s, gives a measure of analgesic potency (Fig. 5, upper graph). In a more sensitive version of the test, the latency of the response is measured (Fig. 5, lower graph). The test can be repeated many times with no change in latency, provided the skin is not damaged by excessive heating. The advantages of the tail flick test include its simplicity and reproducibility, and the fact that the stimulus is quantifiable. Furthermore, the results of the test are well correlated with the analgesic potency of opiate drugs (D'Amour and Smith 1941). Disadvantages are limitation to suitable species and restriction of the region tested to the tail.

The back skin twitch test (Andrews and Workman 1941; Irwin et al. 1951; Winder et al. 1946) is similar and allows an estimate of analgesia over the trunk. Animals used include the rat, guinea pig, and dog. The reflex can be initiated using several different kinds of noxious stimuli, including radiant heat. The intensity of the stimulus is graded, using a fixed duration, rather than measuring a variable latency at a fixed intensity, as in the tail flick test.

The jaw-opening reflex is commonly used in animals that are larger than appropriate for the tail flick or that lack the apparatus. For example, this test can be used in cats, rabbits (Fung et al. 1976; Mitchell 1964; Oliveras et al. 1974b, 1977b, 1979), monkeys (Ha et al. 1978), and humans (Fung et al. 1978; McGrath et al. 1981; Yu et al. 1973). For evoking a jaw-opening reflex, an electrode is im-

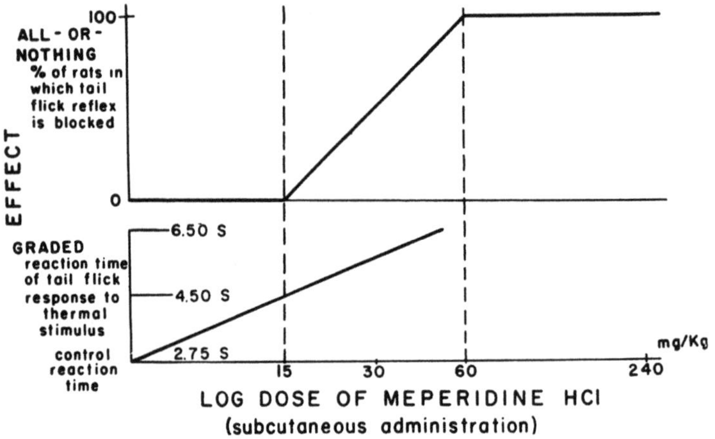

Fig. 5. Two ways to quantitate the analgesic effect of a drug using the tail flick test in the rat (D'Amour-Smith method). At the *top*, the percentage of animals that fails to respond to heating the tail for 20 s is plotted against the dose of the drug. At the *bottom*, the reaction time for a response is plotted against dose. The latter is clearly the more sensitive test. (Grumbach 1966)

planted in or on a tooth and electric currents are passed at increasing intensities until the reflex is elicited. One means of quantitating the response is by recording the electromyogram (EMG) from the masseter muscle (McGrath et al. 1981). During the jaw-opening reflex, EMG activity in the masseter (an extensor muscle) is inhibited. Another way to monitor the jaw-opening reflex is to record the EMG from the digastric muscle (a flexor muscle), which contracts during the reflex (Oliveras et al. 1974b, 1977b; Sessle and Hu 1981; however, cf. Fung et al. 1978). One difficulty with the jaw-opening reflex is that it has two components: a low-threshold one not associated with nociceptive input, and a high-threshold portion due to nociceptive stimulation (McGrath et al. 1981). It is difficult to predict whether or not pain will occur in humans on the basis of the presence or configuration of the reflex inhibition of the masseter (McGrath et al. 1981). Furthermore, narcotic analgesic administration can eliminate the pain without changing the reflex inhibition (McGrath et al. 1981). This is true also for acupuncture analgesia (Fung et al. 1978).

The writhing response of rodents is probably also in part a flexion reflex. This has been used to test for analgesia to visceral pain (Collier and Schneider 1969; Giesler and Liebeskind 1976; Hayashi and Takemori 1971; Hendershot and Forsaith 1959; Siegmund et al. 1957). A noxious chemical substance such as phenylquinone, hypertonic saline, or acetic acid is injected intraperitoneally, and the frequency of writhing of the animal is determined. The movements consist of repeated contractions of the abdominal muscles and extension of the hind limbs. An advantage of this test is that it appears to involve a relatively weak nociceptive stimulus, and so the response can be altered by relatively weak analgesics (Hendershot and Forsaith 1959).

In addition to reflexes involving the contraction or relaxation of the somatic musculature, efforts have been made in the past to measure autonomic reflex responses to noxious stimuli. In general, such tests are now believed to be unreliable, and are therefore not often used in current work.

2.1.3 Other Behavioral Responses

The reflex tests are simple and reliable, but they do not provide a measure of possible changes in higher nervous system function. For testing analgesia due to the latter, a number of approaches have been devised. These include the hot plate test, threshold for vocalization, and shock titration, among others. While several of these tests have a reflex component, they reflect higher-order processing as well.

In the hot plate test (Eddy and Leimbach 1953; Eddy et al. 1950; Johannesson and Woods 1964; Kayan et al. 1969; Woolfe and MacDonald 1944), the animal used is a small rodent. The animal is placed on a hot plate having a temperature in the noxious heat range, and the latency of a behavioral response – consisting of licking of a paw, kicking, or jumping – is measured. This test is very commonly used, often in parallel with the tail flick test (e.g., Hammond et al. 1980; Jacquet and Lajtha 1976; Lewis and Gebhart 1977a,b; Sharpe et al. 1974; Yaksh et al. 1976b; Yaksh and Rudy 1977a; Yaksh and Wilson 1979).

In the vocalization test, a small rodent has its tail clamped or shocked with an electric current (Dandiya and Menon 1963; Dickenson et al. 1979; Haffner 1929). The threshold stimulus necessary for evoked vocalization is recorded.

For the shock titration test, the animal has control of the stimulus intensity. The test can be used in rats or in primates (Oleson et al. 1980; Ross 1966; Schmidek et al. 1971; Weiss and Laties 1958, 1961, 1964; Weitzman and Ross 1962; cf. Boren and Malis 1961; Kelly and Glusman 1968; Pert and Yaksh 1974). The stimulus is generally an electric shock, and the intensity is incremented on repeated trials. The animal can reduce the stimulus intensity at will, and the intensity is kept by the animal just below the threshold for what is assumed to be painful. This test has the advantage that the animal is able to accept only those stimulus intensities it can tolerate, and the time course of changes in the response to the noxious stimulus can be followed.

Many other tests have also been used to demonstrate analgesia in animals (Balagura and Ralph 1973; Dubuisson and Dennis 1977; Evans 1961; Green et al. 1951; Herz et al. 1970; Jacquet and Lajtha 1976; Kitchell 1962; Lewis and Gebhart 1977a; Melzack and Melinkoff 1974; Melzack et al. 1958; Sewell and Spencer 1976; Woolf et al. 1980).

2.1.4 Human Studies

Although human subjects can give verbal reports related to true analgesia, there are many difficulties in human experimentation. One problem is in quantitation. Some of the means that have been developed for quantitating human clinical pain include systematic verbal reports, such as to the McGill Pain Questionnaire and several versions of rating scales, including fixed interval and analogue types (Agnew and Merskey 1976; Houde 1966; Huskisson 1974; Joyce et al. 1975;

Kremer et al. 1981; Loan et al. 1968; Melzack 1975a; Melzack and Torgerson 1971; Ohnhaus and Adler 1975; Revill et al. 1976; Woodfore and Merskey 1972; see also Duncan et al. 1978). An especially interesting development is that it is possible to distinguish between a subject's pain sensation and the unpleasantness of the pain. A problem in work with human patients is that about a third of patients will report a reduction in pain following administration of a placebo (Beecher 1959; Evans 1974; Wilson 1962; Wolf 1959; Wolff and Goodell 1943). Although there is evidence for a physiological mechanism of the placebo effect (Levine et al. 1978), nevertheless this effect can make evaluation of a treatment mode very difficult. Double-blind procedures and careful statistical analysis are generally necessary to sort out effects due to treatment versus those due to the placebo effect (Houde 1966; Houde et al. 1960; Ross et al. 1962; Thorsteinsson et al. 1978). Pain tolerance is a more important consideration than pain threshold (Houde 1966).

A useful approach when painful experimental stimuli are given to a human subject is a psychophysical analysis. Human psychophysics involves the application of known stimuli in a carefully controlled setting, and the measurement of a response. In the case of pain, some forms of stimulation in psychophysical experiments have been electrical shocks, radiant heat, noxious cold, distension of hollow viscera, ischemia, tooth pulp stimulation, and injection of algesic chemicals (Chapman and Jones 1944; Goldberger and Tursky 1976; Gracely et al. 1978a, b, 1979; Hardy 1962; Hardy et al. 1940, 1952; Harrison and Bigelow 1943; Jarvik and Wolff 1962; Lewis 1942; Martin and Chapman 1979; Mayer at al. 1977; Moore et al. 1979; Satran and Goldstein 1973; Smith et al. 1966; Wolf and Hardy 1941; Wolff B. B. 1971, 1978; Wolff and Hardy 1947; Wolff et al. 1940; Wolff and Wolf 1948).

Evaluation of experimental pain can be done using some of the same methods as used for estimates of clinical pain intensity, including verbal descriptions and analogue scales (Gracely et al. 1978a,b; Klepac et al. 1981). Another approach is to record reflex responses as in animal experiments. The human flexion reflex may have a similar threshold to reports of pain (Hardy 1953; Price 1972; Willer 1977), although in the case of the jaw-opening reflex the threshold for the reflex is less than that of pain (McGrath et al. 1981). However, it can be shown that descending control systems and opiate analgesia can, under certain circumstances, affect the flexion reflex and pain differently (McGrath et al. 1981; Willer et al. 1979).

A recently developed statistical approach to the study of human responses to sensory stimuli is based on signal detection theory. The subject is submitted to stimuli of various intensities, and he is asked to signify when a suprathreshold stimulus occurs. The subject can correctly indicate that a stimulus occurred, he can correctly state that a stimulus did not occur, he can incorrectly signal a stimulus that did not occur, or he can fail to detect a stimulus that did occur. It is believed that the subject establishes a criterion to enable him to recognize occurrence of a stimulus above a noise level. The criterion can be shifted to allow a greater proportion of stimuli to be recognized, but in this case there will be more false positive responses, as well, because noise will exceed the criterion level more often. By a suitable analysis, it is thought possible to distinguish between responses to sensory information and changes due to alterations in the criteria used by the subject to identify stimuli. The signal detection method has been used for the study of experimental pain by several groups (Chapman et al. 1975a, b, 1977; Clark 1969; Clark and Mehl 1971; Clark and Yang 1974; Lloyd and Wagner 1976). It is interesting that a placebo can be shown to alter the willingness of a subject to report pain (change in response criterion), but not the sensitivity to pain (Clark 1969; Feather et al. 1972). However, this approach has been criticized (McBurney 1975, 1976; McCreery and Bloedel 1978; Rollman 1977, 1979; cf. rebuttals by Chapman 1977 and Jones 1979).

2.1.5 Neural Correlates of Analgesia

A major effort in the study of nociceptive mechanisms is directed at a determination of the neural correlates of nociception and antinociception. Investigations of this type may involve the study of neural activity in awake, behaving animals or in reduced, often anesthetized preparations. There are advantages and disadvantages associated with each approach.

Recordings from neurons in awake, behaving animals approach investigations on human subjects in terms of studying the activity of the nervous system with minimal interference by drugs or surgery. However, experimental interventions are difficult in such preparations, and it is very hard to develop experimental paradigms that permit simple interpretation of the data. For example, in addition to responses to nociceptive inputs, central nociceptive neurons may show alterations in their discharges that reflect the behavioral relevance of the stimuli (Dubner et al. 1981; Hayes et al. 1979b, 1981).

On the other hand, work on reduced , particularly on anesthetized, preparations may be misleading in terms of neural activity that relates to nociception in the awake, behaving animal. Nevertheless, insofar as recordings from neurons in anesthetized animals reflect nociceptive mechanisms, progress in understanding neural connectivity and synaptic mechanisms is likely to be much more rapid under the controlled conditions possible in such preparations.

Sections 4 and 5 will be concerned with studies on neural correlates of analgesia, with emphasis on pathways descending from the brain that modulate the flexion reflex and the nociceptive responses of spinal cord neurons.

2.1.6 Analgesia Tests in Studies of Pain Mechanisms

The remainder of this section will be used to consider various forms of experimental manipulations that have been reported to cause "analgesia". In most cases, the "analgesia" was observed in animal subjects and was judged on the basis of alterations in one or more of the tests for analgesia just described. In some cases, a similar analgesia has been found in human subjects, confirming the impression of analgesia in the animal subjects. One goal of such studies is to understand the neural systems underlying pain and pain control.

2.2 Stimulation-Produced Analgesia

During the 1960s there was a considerable interest in the observation that anesthesia or analgesia could be produced by passing electric currents through the brain (Reynolds and Sjoberg 1971)[1]. In an attempt to determine the neuroanatomical site(s) responsible for the anesthetic effects of electrical stimulation of the whole brain, Reynolds (1969) introduced a focal stimulating electrode into the midbrain and found that there are sites within the periaqueductal gray (PAG) that when stimulated produce a profound analgesia. During such stimulation, Reynolds could perform a laparotomy on the rats without their showing any signs of discomfort, although the rats were obviously awake, had intact motor control, and responded in an apparently normal fashion to other, innocuous, stimuli. This phenomenon was named "stimulation-produced analgesia" (SPA) (Mayer and Liebeskind 1974).

Since Reynolds' study, a number of investigators have confirmed his observation that SPA can result from stimulation of the PAG; furthermore, many other sites within the brain stem are also effective, especially sites near the midline, from the level of the diencephalon caudally to the medulla (Fig. 6; Akaike et al. 1978; Balagura and Ralph 1973; Basbaum et al. 1977; Bennett and Mayer 1979; Cox and Valenstein 1965; Dubuisson and Dennis 1977; Gebhart and Toleikis 1978; Giesler and Liebeskind 1976; Gol 1967; Goodman and Holcombe 1976; Hayes et al. 1979c; Liebeskind et al. 1973; Mayer and Liebeskind 1974; Mayer et al. 1971; Melzack and Melinkoff 1974; Mohrland and Gebhart 1980; Oleson et al. 1978, 1980; Oliveras et al. 1974a,b, 1975, 1979; Rhodes and Liebeskind 1978; Sandberg and Segal 1978; Satoh et al. 1980a; Schmidek et al. 1971; Segal and Sandberg 1977; Sessle and Hu 1981; Soper 1976; Yeung et al. 1977; Zorman et al. 1981). Besides the PAG and the periventricular gray (PVG), particularly important sites for SPA (see below) are the nucleus raphe magnus (NRM) (Besson et al. 1981; Oleson et al. 1978, 1980; Oliveras et al. 1979; Proudfit and Anderson 1975; Satoh et al. 1980a; Sessle and Hu 1981; Zorman et al. 1981) and the nucleus paragigantocellularis (or magnocellularis) of the reticular formation (Akaike et al. 1978; Satoh et al. 1980a; Zorman et al. 1981). The reason for the importance of these sites for SPA is that neurons in the NRM and the nucleus paragigantocellularis project directly to the spinal cord. Descending influences from the PAG and PVG are thought to be, for the most part, relayed through synapses in the lower brain stem. Figure 6 shows the distribution of some of the points from which SPA can result in cats. Other brain regions, including the cerebellum (Oleson et al. 1980; Siegel and Wepsic 1974), ventrobasal thalamus (Boethius et al. 1976; Hosobuchi et al. 1973; Mazars 1975; Mazars et al. 1973, 1979), internal capsule (Adams et al. 1974; Fields and Adams 1974); lateral hypothalamus (Balagura and Ralph 1973; Cox and Valenstein 1965; Yunger et al. 1973), septum and preoptic area (Breglio et al. 1970; Gol 1967; Oleson et al. 1980; Schmidek et al. 1971), and

1 An impressive experimental attempt to determine the effect of electrical stimulation on the brain can be attributed to Volta. According to Stevens (1970), "In the year 1800 Volta assembled a large battery of his newly invented cells and connected the total array to a pair of metal rods inserted in his ears. When he closed the switch he felt a jolt in the head, followed by a noise like the boiling of thick soup"

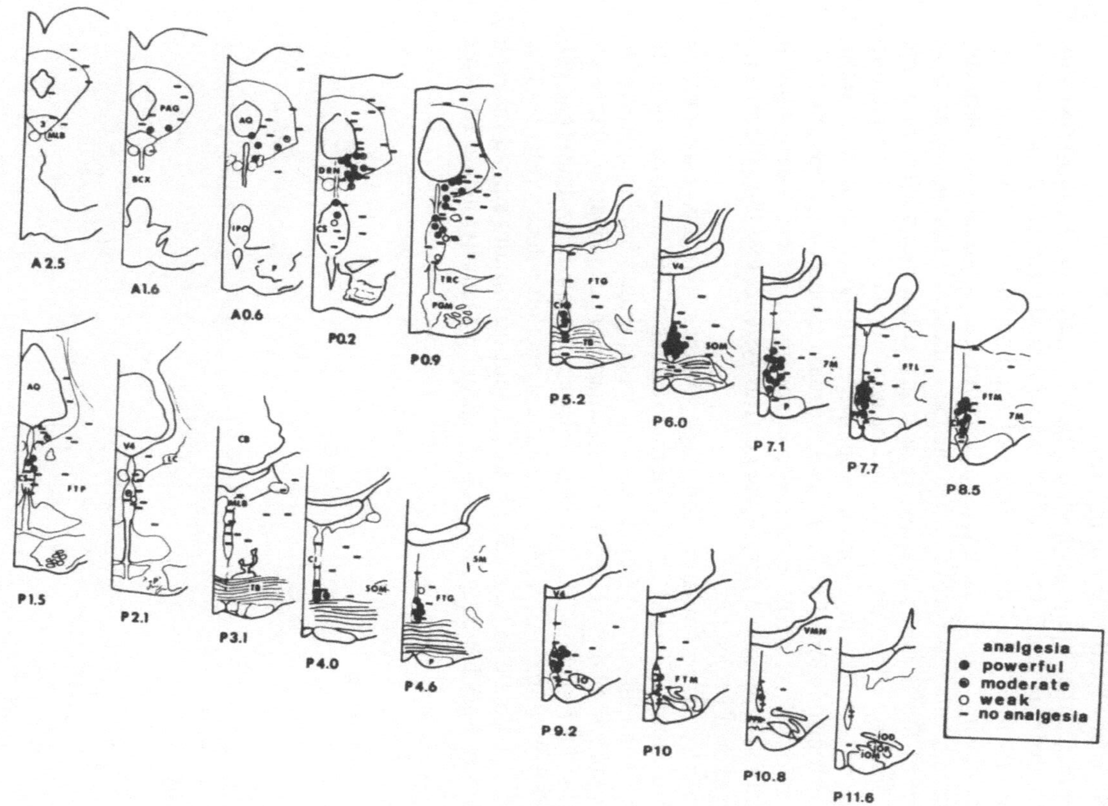

Fig. 6. Sites in the brain stem of the cat that when stimulated resulted in stimulation-produced analgesia. The animals were awake and behaving. The pinch test was employed. (Oliveras et al. 1979). *CS*, central superior; *CI*, central inferior; *PAG*, periaqueductal gray

caudate nucleus (Lineberry and Vierck 1975; Schmidek et al. 1971) have also been reported to be effective in causing SPA.

It should be stressed that stimulation within the CNS can lead to actions that may be difficult to attribute to activation of particular neural elements in the absence of other, parallel lines of evidence. Some of the hazards of CNS stimulation include the following. Given an accurate placement of a stimulating electrode within a particular nucleus (proven after the experiment by histology), an electrical stimulus of low intensity may excite neurons within that nucleus or axons of passage. Thus any effects observed could be attributed either to the excitation of the nucleus or to volleys in the axons of passage. The latter could have effects by orthograde transmission of volleys down the main axonal projection, by antidromic activation of recurrent collaterals, or by spread of excitation through axon collaterals synapsing in a part of the nervous system other than that receiving the main axonal projection.

In the case of electrical stimulation with high stimulus strengths, care must be taken to avoid stimulus spread to neighboring regions of the CNS. An estimate of the possible stimulus spread from a unipolar metal gross stimulating electrode to nearby myelinated fibers is given in recent papers as approximately 10 μm/μA (Abzug et al. 1974; Ranck 1975). However, this relationship is nonlinear and will vary for different conditions, such as axonal conduction velocity (Fig. 7). It is advisible for an estimate to be made of the amount of stimulus spread for each set of conditions. Paradoxically, with a high stimulus intensity, neural elements

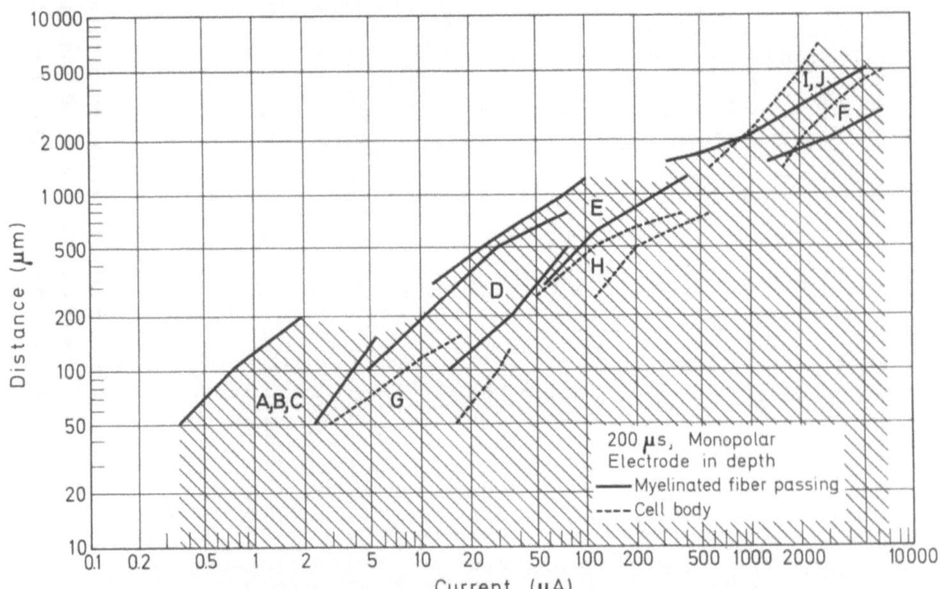

Fig. 7. Estimates of stimulus spread. The graph shows the relationship between stimulus intensity (μA; 200-μs pulse) and distance from the electrode to the neural elements stimulated. The letters designate the different studies from which the data were derived. Myelinated axons are the neural elements involved for experiments indicated by solid lines and cell bodies when dotted lines are used. Note that a stimulus of, say, 10 μA excites elements at a distance of 100 – 200 μm; 100 μA excites elements at 1000 μm. (Ranck 1975)

nearest the electrode may not be activated, whereas elements at a greater distance may (Ranck 1975).

One way to excite nearby neurons, but not axons, is by the release of excitatory agents, such as glutamate ions, that act preferentially on receptors on the cell body, and on dendrites, but not on axons (Behbehani and Fields 1979).

Commonly employed measures of analgesia are flexion reflexes, such as the tail flick and the jaw-opening reflexes. Other, more complex, behavioral measures have also been used. By measuring the extent and duration of alterations in such tests as the tail flick, it is possible to determine the stimulus parameters needed for SPA and to describe some of the features of the analgesia. For example, the stimulus parameters used for SPA have varied widely across studies (Table 1). It is not certain that all of these studies avoided the problem of stimulus spread (cf. Fig. 7). However, strong stimulation may be required to activate small neural elements in a sufficient volume of tissue to produce a behavioral change. Another factor in animals having chronically implanted electrodes is tissue damage around the tip of the stimulating electrode (Zorman et al. 1981).

Analgesia often outlasts stimulation in the rat by periods of several minutes (Mayer and Liebeskind 1974; Mayer et al. 1971; Reynolds 1969; Rhodes and Liebeskind 1978). By contrast, it was initially reported that SPA rarely outlasts stimulation in the cat by more than a few seconds (Liebeskind et al. 1973; Oliveras et al. 1974a, 1975). Thus in the rat it is easy to test for analgesia after other stimulus-locked events, such as aversive reactions, have terminated, whereas this has not been feasible in several studies in the cat. However, this difference may be due to variations in stimulus duration or intensity, rather than to species, since weaker stimuli in the rat do not produce analgesia that outlasts stimulation for very long (Yeung et al. 1977), and prolonged stimulation in the cat results in analgesia that outlasts stimulation (Melzack and Melinkoff 1974; Oliveras et al. 1979; see also Dubuisson and Dennis 1977). Stimulation-produced analgesia outlasts strong stimulation in the monkey PAG by up to 5 min (Hayes et al. 1979c), and SPA from medial diencephalic loci can be quite prolonged in the monkey, lasting 2 h after stimulation (Oleson et al. 1980).

The onset of SPA may be gradual, requiring several minutes for its full expression (Melzack and Melinkoff 1974), or it may be immediate (Oleson et al. 1980). As mentioned earlier, SPA is not accompanied by overt changes in motor control, responses to other sensory stimuli (visual, auditory, tactile), or emotion, attention, or motivation (Mayer et al. 1971; Mayer and Liebeskind 1974; Melzack and Melinkoff 1974; Soper 1976; Oliveras et al. 1975; Yeung et al. 1977; Yunger et al. 1973). The hypothesis that activation of "reward" mechanisms can account for SPA (Cox and Valenstein 1965; Gol 1967) does not appear to be a viable one, since self-stimulation and "analgesia" can be produced independently (Balagura and Ralph 1973; Breglio et al. 1970; Mayer and Liebeskind 1974; Sandberg and Segal 1978; Yunger et al. 1973). Stimulation-produced analgesia, therefore, appears to depend upon a relatively specific neural mechanism.

Although in some studies there appeared to be a relationship between the brain stem site causing SPA and the peripheral distribution of the analgesia (Mayer and Liebeskind 1974; Mayer et al. 1971), in other studies there was no clear-cut somatotopic relationship, the analgesia being either generalized over much of the body surface or localized in a way not predictable from the particular stimulus

Table 1. Parameters for stimulation-produced analgesia

Subject	Site(s)	Train	Freq. (Hz)	Intensity	Reference
Rat	PAG[a]	Continuous	60	15 – 30 μA rms	Reynolds 1969
	PAG[a], PVG[b]	Repeated 100 ms trains at 2 – 3/s for 15 – 60 s	40 – 100	5 – 250 μA rms or 250 – 8000 μA	Mayer et al. 1971; Mayer and Liebeskind 1974; Mayer and Hayes 1975; Akil et al. 1976; Yaksh et al. 1976b; Rhodes and Liebeskind 1978; Hayes et al. 1977a
	PAG[a], NR paragig.[c]	Continuous	50	< 10 μA in anesthetized and 30 – 80 μA in chronic animals	Zorman et al. 1981
	NR paragig.[c]	30 – 60 s	20	50 – 200 μA	Akaike et al. 1978
Cat	PAG[a], Raphe	Repeated 200 ms trains at 3/s for 10 – 60 s	50 – 100	5 – 100 μA or 2 – 10 V	Liebeskind et al. 1973; Oliveras et al. 1974a; 1979
	PAG[a]	Continuous	100	25 – 50 μA	Gebhart and Toleikis 1978
	PAG[a] CTegT[d]	500 ms trains at 2/s	100	350 – 400 μA	Melzack and Melinkoff 1974
Monkey	PAG[a]	10 – 60 s	20	2 – 5 mA	Hayes et al. 1979b
	PVG[b], Other	1 – 30 min	20	0.5 – 9 mA	Oleson et al. 1980
Human	PVG[b]	Continuous	10 – 20	3 – 6 V	Hosobuchi et al. 1977
	PVG[b], PAG[a]	Continuous	25 – 75	< 5 V	Richardson and Akil 1977a
	VPM[e]	Continuous	100 – 125	1 – 1.5 mA	Hosobuchi et al. 1973

[a] Periaqueductal gray
[b] Periventricular gray
[c] Nucleus reticularis paragigantocellularis
[d] Central tegmental tract
[e] Ventral posterior medial nucleus

location (Lewis and Gebhart 1977a,b; Rhodes and Liebeskind 1978; Yeung et al. 1977).

Evidence that there may be neural systems that when active produce analgesia can come not only from studies in which neural tissue is stimulated but also from lesion studies. For instance, Melzack et al. (1958) reported that animals with bilateral lesions in the region of the central tegmental tract were hyperalgesic (cf. Kelly and Glusman 1968). This observation would be consistent with interruption of an analgesia-producing pathway. However, animals with lesions of the PAG show either a reduced responsiveness to noxious stimuli (Halpern 1968; Melzack et al. 1958; Skultety 1958) or no change (Kelly and Glusman 1968). These findings are difficult to reconcile with the results of experiments on SPA, unless the PAG is regarded as a region that is involved in both analgesia and aversive responses (cf. Nashold et al. 1969; Pert and Walter 1976; Schmitt et al. 1974; Skultety 1963), and so lesions in the PAG can result in a cancellation of opposing effects. Another explanation would be that the PAG is not tonically active, so that a lesion might not produce a change. Lesions of the NRM or injections of local anesthetic into this nucleus, on the other hand, produce hyperalgesia (Proudfit 1980; Proudfit and Anderson 1975).

Stimulation-produced analgesia has also been described in humans (Sweet 1977). Patients with chronic pain not treatable by other means have been implanted with electrodes in the PVG (Fig. 8) or PAG. Stimulation in these areas has in some cases been successful in relieving the clinical pain state (Adams 1976; Boëthius et al. 1976; Hosobuchi et al. 1977a; Richardson and Akil 1977a,b; however, cf. Gybels 1979; Mazars et al. 1979). In the human cases, the pain is often relieved for 3 – 4 h following a 30-min period of stimulation (Hosobuchi et al. 1977a). Stimulus parameters for successful pain relief are set by the patient, and fall into the range shown in Table 1. Generally, the PVG is a better target than the PAG because of fewer side effects and a more potent analgesia. Stimulation in the ventrobasal complex or in the internal capsule is effective for pain

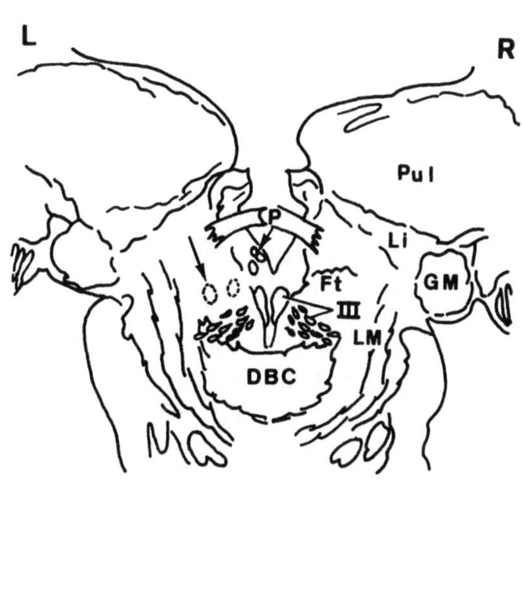

Fig. 8. Site for stimulation-produced analgesia in the human. The *medial arrow* points to a site in the periventricular gray that was effective when stimulated electrically in relieving pain from carcinoma of the rectum. Another more lateral site is shown by a *second arrow*. This site was ineffective. (Hosobuchi et al. 1977a). *DBC*, decussation brachium conjunctivum; *Ft*, reticular formation; *GM*, medial geniculate body; *Li*, nucleus limitans; *LM*, medial lemniscus; *P*, posterior commissure; *Pul*, Pulvinar; *III*, oculomotor nucleus and nerve

originating within the CNS, as in the thalamic syndrome or after injury to the cerebral cortex (Adams et al. 1974; Boëthius et al. 1976; Fields and Adams 1974; Hosobuchi et al. 1973, 1974; Mazars 1975; Mazars et al. 1973, 1979). Analgesia from stimulation in these areas is often accompanied by paresthesias (Boëthius et al. 1976; Fields and Adams 1974; Hosobuchi et al. 1973; however, cf. Mazars et al. 1979), whereas paresthesias are not a feature of SPA from periventricular electrode placements.

2.3 Opiate and Opioid Analgesia

Since ancient times, opiate drugs, such as opium and morphine, have been used to relieve pain (Reynolds and Randall 1957). Why such complex substances manufactured by plants should have a potent action in humans was unclear (Collier 1972) until recently. First, it was found that there are specific receptors for opiates in nervous tissue (Hiller et al. 1973; Pert and Snyder 1973; Simon et al. 1973; Terenius 1973a,b). Recently, opiate receptors have been subdivided into at least two different classes, μ and δ (Lord et al. 1977), based on their relative sensitivity profiles to different ligands (Audigier et al. 1980; Chang and Cuatrecasas 1979; Chang et al. 1979; Della Bella et al. 1978; Gillan et al. 1979; Lord et al. 1977; Snyder and Goodman 1980; Terenius 1977; cf. Gilbert and Martin 1976; Martin W. R. et al. 1976). The relative abundance of μ and δ opiate receptors varies in different brain regions (Chang et al. 1979; Ninkovic et al. 1981). Morphine-like agents bind better to μ receptors and enkephalin-like substances bind better to δ receptors. Analgesia is better correlated with binding to μ than to δ receptors (Audigier et al. 1980; cf. Urca et al. 1978).

Once it became evident that there are opiate receptors, a search was made for endogenous ligands that might act on these receptors (Kosterlitz 1978). Soon, two pentapeptides, methionine- and leucine-enkephalin (Fig. 9A,B), were discovered in the brain of the pig (Hughes 1975; Hughes et al. 1975). The same substances were then found also in cow brain (Simantov and Snyder 1976c). These pentapeptides can be released by stimulation of nervous tissue; they interact with opiate receptors and their administration causes analgesia. Their action can be blocked by naloxone, and they produce cross-tolerance with morphine (Belluzzi et al. 1976; Buscher et al. 1976; Malick and Goldstein 1977; Puig et al. 1977; Simantov and Snyder 1976a,b; Takagi et al. 1978; Waterfield et al. 1976; Yaksh et al. 1977b). Enkephalins are degraded by peptidases, and so their potency and duration of action are limited under ordinary in vivo conditions. However, analogues have been synthesized that resist degradation, and these analogues are more potent and have longer durations of action than the enkephalins (Bajusz et al. 1976; Dutta et al. 1977; Roemer et al. 1977; Stacher et al. 1979; Walker et al. 1977; Yamashiro et al. 1977; however, cf. Von Graffenried et al. 1978). It was proposed that the enkephalins might serve as neurotransmitters or as modulators (Kosterlitz and Hughes 1975). The two enkephalins may be contained in separate populations of neurons (Larsson et al. 1979; however, see Watson et al. 1977).

Other opioid substances in addition to the enkephalins have been found in the brain or the pituitary; these include long-chain peptides such as β-endorphin

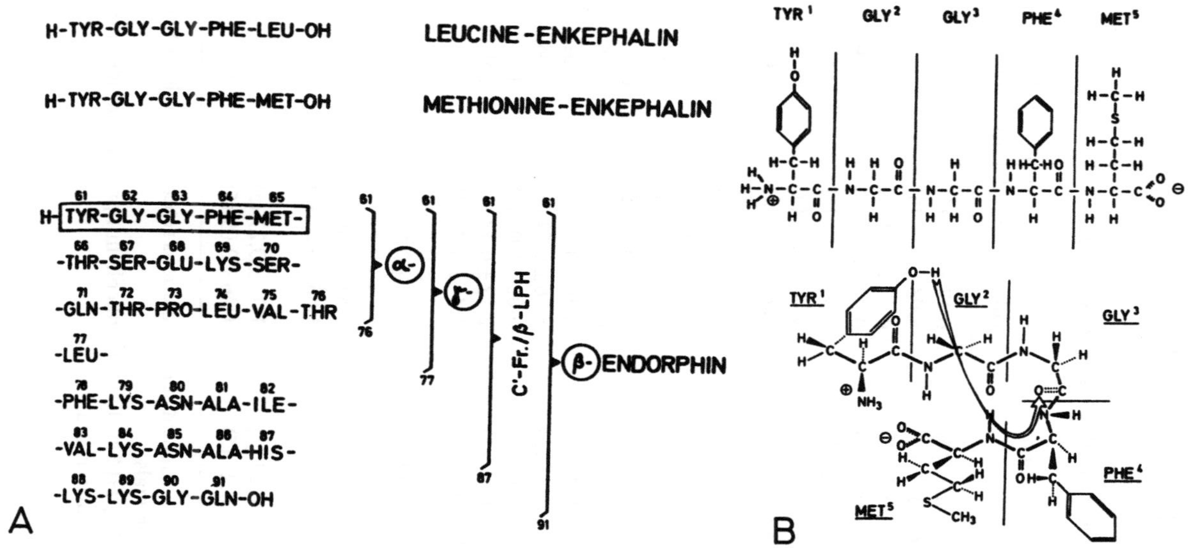

Fig. 9. In **A** are shown the amino acid sequence of the enkephalins and several endorphins. The numbers above the amino acids are the positions within the β-lipotropin pituitary hormone. In **B** are the structural formula of methionine-enkephalin *above*, and a representation of the possible β-bend configuration of the molecule stabilized by a hydrogen bond between the CO group of the Gly^2 residue and the NH group of the Met^5 residue. The *arrow* indicates the location of an additional hydrogen bond. (Teschemaker 1978)

Table 2. Distribution of opiate binding sites in bovine brain and spinal cord (Ninkovic et al. 1981)

	Specific binding[a] (fmol/mg protein)		
	^{3}H-Morphine (0.5 nm)	D-[^{3}H]Ala^2,D-leu^5-enkephalin (0.5 nm)	Binding ratio[b]
Putamen	8.62±0.53 (6)	6.17±0.62 (6)	1.4
Caudate	12.96±0.55 (7)	8.08±0.37 (6)	1.6
Globus pallidus	4.4 ±0.55 (5)	4.5 ±0.63 (5)	0.98
Substantia nigra	3.32±0.33 (5)	0.9 ±0.2 (5)	3.69
Frontal cortex	3.46±0.33 (5)	6.85±0.4 (5)	0.5
Parietal cortex	2.86±0.4 (5)	4.08±0.64 (5)	0.71
Pyriform cortex	7.76±0.36 (5)	4.92±0.12 (5)	1.6
Thalamus	7.12±1.11 (5)	3.4 ±0.17 (5)	2.1
Midline thalamus	7.1 ±0.5 (5)	3 ±0.47 (5)	2.37
Periaqueductal gray	8.28±0.72 (5)	4.46±0.22 (5)	1.85
Raphe nucleus	4.8 ±0.47 (5)	2.7 ±0.21 (5)	1.8
Hippocampus	3.66±0.45 (5)	9.24±0.47 (5)	0.4
Dentate gyrus	2.3 ±0.43 (5)	6.47±0.71 (5)	0.36
Hypothalamus	3.78±0.47 (5)	2.34±0.4 (5)	1.61
Amygdala	9 ±1.17 (5)	5.36±0.57 (5)	1.68
Nucleus accumbens	8.26±0.87 (5)	6.02±1.09 (5)	1.37
Superior colliculi	5.16±0.67 (5)	2.72±0.48 (5)	1.9
Lateral geniculate	1.88±0.2 (5)	1.4 ±0.28 (5)	1.34
Olfactory bulb	2.68±0.31 (5)	2.7 ±0.43 (5)	1
Olfactory tubercle	4.66±0.39 (5)	2.48±0.26 (5)	1.92
Cerebellum	2.48±0.36 (5)	1.54±0.17 (5)	1.61
Medulla	2.02±0.56 (5)	1.53±0.35 (5)	1.32
Pons	1.76±0.33 (5)	1.88±0.27 (5)	0.93
Spinal cord –			
Dorsal horn (gray)	3.07±0.24 (5)	1.7 ±0.27 (5)	1.8
Ventral horn (gray)	1.92±0.2 (5)	1.46±0.09 (5)	1.31

[a] Specific binding represents the mean of triplicate determinations repeated on at least five separate occasions using different animals

[b] The binding ratio ([^{3}H]morphine/D-[^{3}H]Ala^2,D-Leu^5-enkephalin) provides an index of the ratio μ to δ binding sites in the various brain areas

(Fig. 9A) and dynorphin (Bradbury et al. 1976; Cox et al. 1975, 1976; Feldberg and Smyth 1977; Goldstein 1976; Goldstein et al. 1979; Guillemin et al. 1977a,b; Li and Chung 1976; Li et al. 1976; Loh et al. 1976; Mains et al. 1977; Pasternak et al. 1975; Rossier et al. 1980; Rubinstein et al. 1977; Terenius and Wahlström 1975; Teschemacher et al. 1975; Tseng et al. 1976; Wahlström et al. 1977), and the dipeptide Kyotorphin (tyrosyl-arginine) (Satoh et al. 1980b; Takagi et al. 1979).

It is interesting that methionine-enkephalin forms a component of β-endorphin and that leucine-enkephalin is a part of dynorphin. The pituitary peptide β-lipotropin appears to be a prohormone for β-endorphin (Cox et al. 1976; Guillemin

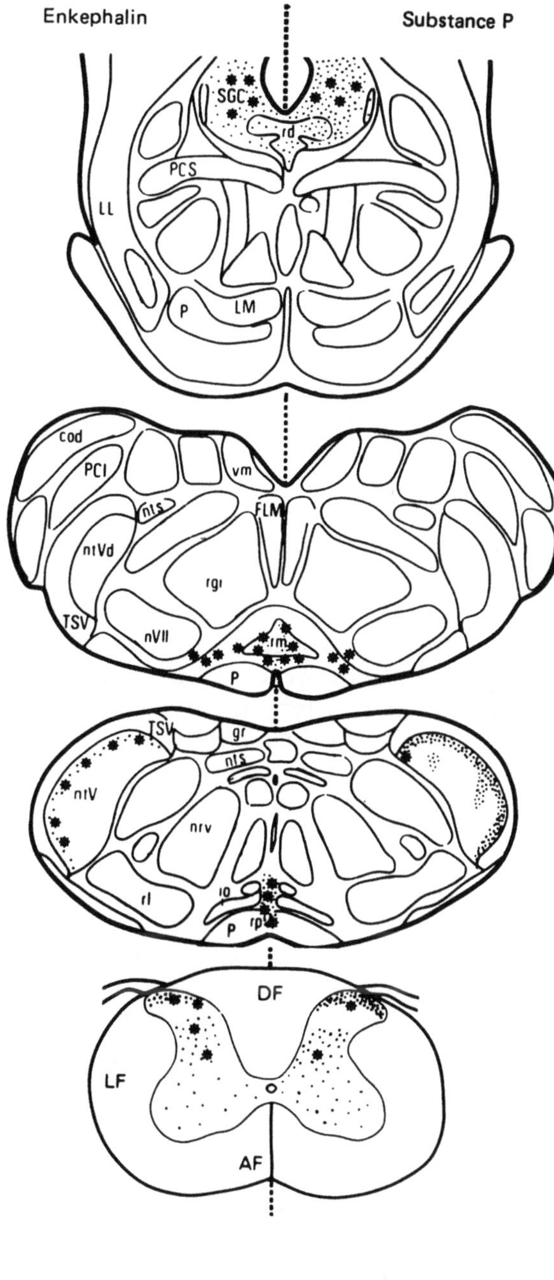

Fig. 10. Distribution of nerve terminals and cell bodies containing enkephalin-like immunoreactivity (*left*) or substance P (*right*). The terminals are indicated by *dots* and the cell bodies by asterisks. (Hökfelt et al. 1977b). *AF*, anterior funiculus; *DF*, dorsal funiculus; *cod*, dorsal cochlear nucleus; *gr*, nucleus gracilis; *LF*, lateral funiculus; *LL*, lateral lemniscus; *LM*, medial lemniscus; *nrv*, nucleus reticularis ventralis; *nts*, nucleus tractus solitarius; *ntV*, spinal nucleus of trigeminal; *nVII*, facial nucleus; *rd*, raphe dorsalis; *rm*, raphe magnus; *rp*, raphe pallidus; *SGC*, periaqueductal gray

et al. 1977a; Lazarus et al. 1976; Loh et al. 1976). It is unclear as yet whether the β-endorphin and dynorphin serve as prohormones for the enkephalins or if the enkephalins are breakdown products of the larger compounds (Beaumont and Hughes 1979; Cox et al. 1975; Lazarus et al. 1976). The brain enkephalin content is not derived from the pituitary, since enkephalin distribution is not changed by hypophysectomy (Cheung and Goldstein 1976; Hong et al. 1977; Kobayashi et al. 1978; Wesche et al. 1977).

β-Endorphin has a greater analgesic potency than either morphine or methionine-enkephalin (Loh et al. 1976), suggesting that it may have a more significant physiological role than methionine-enkephalin. Furthermore, the very high potency of dynorphin (700 times more potent than leucine-enkephalin and 50 times more potent than β-endorphin) leads to the proposal that this substance may be a physiologically active neurotransmitter (Goldstein et al. 1979). It may well turn out that several different opioid substances are neurotransmitters or modulators. This notion is supported by studies showing that separate neuronal populations contain peptides of different lengths. Met-enkephalin-containing neurons are present throughout the CNS, whereas β-endorphin-containing neurons are in a restricted region of the hypothalamus (Bloom et al. 1978; Watson et al. 1978).

The endogenous opioids and the opiate receptors are concentrated in regions of the CNS (Table 2; Fig. 10) that take part in SPA or are involved in nociceptive transmission (Aronin et al. 1981; Atweh and Kuhar 1977a,b,c; Dupont et al. 1980; Elde et al. 1976; Fields et al. 1980; Hökfelt et al. 1977a,b, 1979; Hong et al. 1977; Hughes et al. 1977; Hunt et al. 1980; Kuhar et al. 1973; LaMotte et al. 1976; Lindberg et al. 1979; Pert et al. 1975; Sar et al. 1978; Simantov and Snyder 1976a; Simantov et al. 1976, 1977; Uhl et al. 1979; Watson et al. 1977; Yang et al. 1977). It has thus been of considerable interest to draw parallels between the neural systems responsible for SPA and for opiate analgesia (Sherman and Liebeskind 1980; see Section 3). However, it should be kept in mind that the endogenous opiate systems are also associated with many brain structures not known to be related to pain mechanisms or known to be involved in other functions, and that there are likely to be systems responsible for SPA that do not involve opioid substances.

2.4 Analgesia from Peripheral Stimulation

Analgesia can be produced not only by stimulation within the brain or by the administration of opiate and other drugs, but also by peripheral stimulation. This has been recognized for many years and has been described as the effect of "counterirritation" (Gammon and Starr 1941). Interestingly, there is evidence that at least some forms of analgesia produced by peripheral stimulation involve the activation of descending control systems utilizing an endogenous opiate mechanism.

One means of producing analgesia by peripheral stimulation is the ancient Chinese therapeutic procedure known as acupuncture (from Latin *acus*, needle, and *punctura*, puncture). This procedure is described in the *Yellow Emperor's Canon of Internal Medicine,* which dates back to several centuries B.C. and which has been attributed to Huang Ti, the "Yellow Emperor," who lived in 2700 B.C. (Bonica 1974; Kao 1973; Lu and Needham 1980). Although these are not pertinent to the present discussion, acupuncture is thought to have effects on many other systems in addition to the nervous system (O'Connor and Benskey 1975).

Acupuncture involves the insertion of needles into appropriate loci on the body (Lu and Needham 1980). Loci are selected on the basis of the organ thought to be

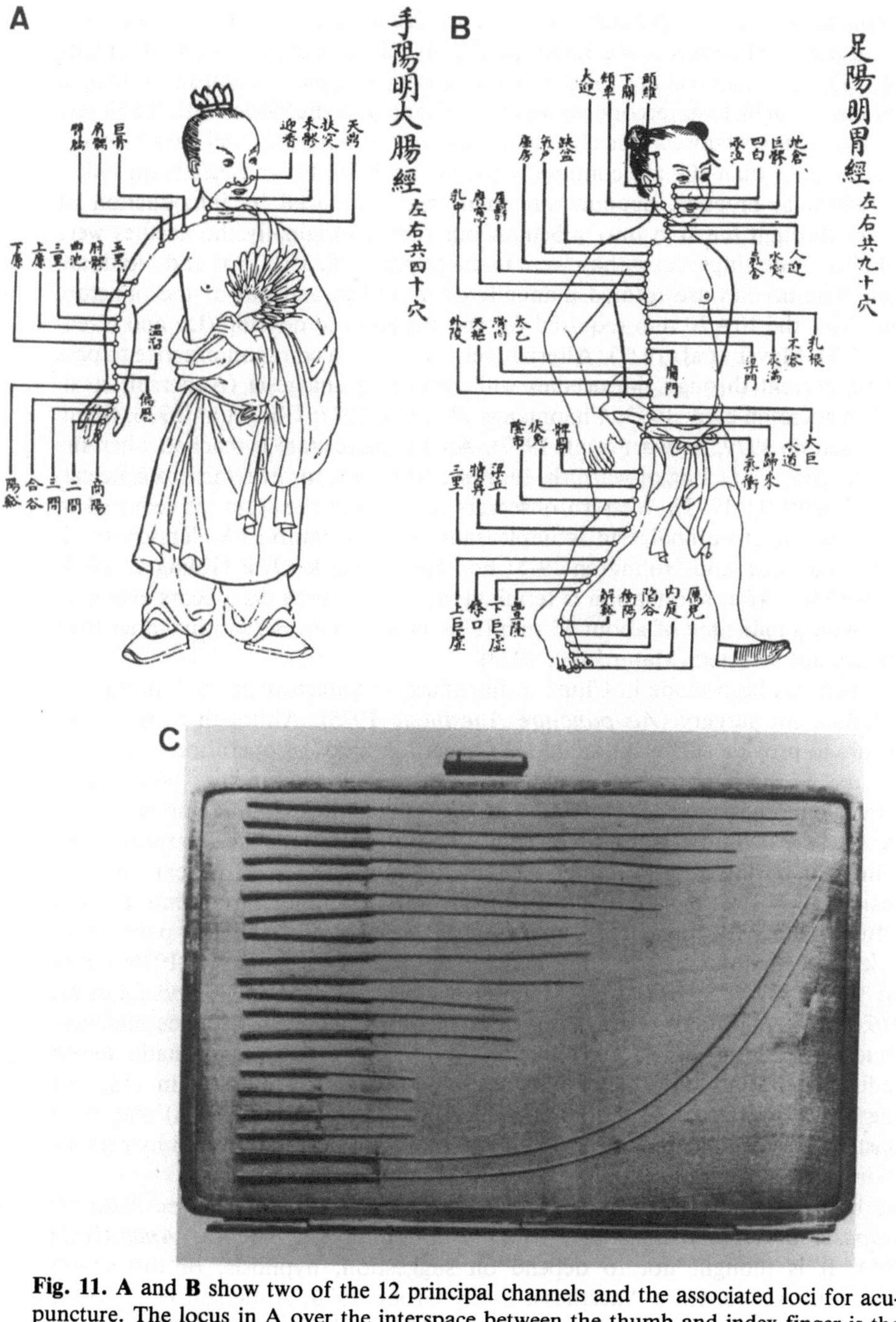

Fig. 11. A and **B** show two of the 12 principal channels and the associated loci for acupuncture. The locus in A over the interspace between the thumb and index finger is the *Ho-ku* point. **C** shows a modern set of acupuncture needles. (Lu and Needham 1980)

the source of a medical problem. The loci are distributed along 12 principal (and several minor) channels or meridians (*Ching-lo*) dividing the body longitudinally (Fig. 11). The channels are thought to be energy conductors and the principal ones are thought to interconnect visceral organs and the body wall. There are also anastomoses between the channels. Although hundreds of loci have been described, fewer than 100 are commonly used (Kao 1973; Lu and Needham 1980). In traditional practice, acupuncture analgesia is induced by introduction of needles through the skin into subcutaneous tissue. Originally the needles were probably stone chips; later they were metal (Nghi et al. 1973; Lu and Needham 1980). The needles are rotated manually or moved in and out of the puncture wound for the 10–20 min required for the analgesia to develop (Lu and Needham 1980; Mayer at al. 1977). Alternatively, it is now a common practice to pass electric current through the needles while inducing analgesia (Anderson et al. 1974; Andersson et al. 1973; Chapman et al. 1976, 1977; Ghia et al. 1976; Mann 1974; Sodipo 1979; Stewart et al. 1977). An analgesic state is reached when the subject reports a feeling of warmth, fullness, numbness, or heaviness, a state called *Teh Ch'i* (Li 1973). Western observers report that the stimulus required to induce acupuncture analgesia is unpleasant or even painful (Andersson et al. 1973; Andersson and Holmgren 1975; Fox and Melzack 1976; Holmgren 1975; Mann 1974). After stimulation is terminated, the analgesia disappears exponentially with a half-time of about 15 min (Han et al. 1980). Sensations other than pain are not affected (Mann 1972, 1974).

An effort has been made in China to introduce acupuncture as an "anesthetic" procedure for surgery (*Acupuncture Anesthesia* 1975). Although acupuncture appears to provide sufficient analgesia for certain kinds of operation in selected patients, Western observers would not predict that acupuncture will replace chemical anesthesia in the near future in Western countries (American Acupuncture Anesthesia Study Group 1976; Taub 1976). In fact, patients who seem not to be in pain during surgery under acupuncture anesthesia can be experiencing considerable pain (Modell et al. 1976). However, acupuncture seems to be a useful method for producing analgesia in a variety of clinical pain states (Eriksson and Sjölund 1976; Ghia et al. 1976; Lee P.K.Y. et al 1975, 1976; Levitt and Walker 1975; Mann 1972; Mann et al. 1973; Sodipo 1979; Spoerel et al. 1976; however, cf. Gaw et al. 1975). Complications can occur with acupuncture, including infection, hematoma, pneumothorax, and cardiac tamponade due to needle penetration; and failure to use other forms of therapy in cases of progressive disorders like acute appendicitis or cancer (Bonica 1974) may have disastrous results. Acupuncture should therefore not be administered by persons lacking suitable anatomical and medical training.

The basis for the effectiveness of acupuncture is controversial (see *National Symposia of Acupuncture and Moxibustion and Acupuncture Anaesthesia* 1979). It is thought not to depend on suggestion, hypnosis, or distraction (American Acupuncture Anesthesia Study Group 1976; Anderson et al. 1974; Andersson and Holmgren 1975; Berlin et al. 1975; Chiang et al. 1973; Croze et al. 1976; Goldberger and Tursky 1976; Kaada 1974; Stewart et al. 1977; however, see Brennan et al. 1973; Day et al. 1975; Kroger 1973). Signal detection studies suggest that in the case of painful stimulation of the skin, acupuncture produces a change in response criterion (Clark and Yang 1974; however, see Chapman et al. 1977; Hayes et al. 1975), whereas with tooth pulp stimulation as the test of pain, acupuncture results in a reduction in sensory sensitivity as well (Chapman et al. 1977; cf. Andersson et al. 1973; Mumford and Bowsher 1973). The traditional Chinese medical theory is that acupuncture alters the distribution of energy (*Chi*) in the body (Lu and Needham 1980; Nghi et al. 1973), which is desirable since many diseases are considered to result from an excess or deficiency of energy. Western and also many Chinese medical scientists are more likely to subscribe to the rival theory that the mechanism of acupuncture depends upon the activation of afferent nerve fibers by the needles and subsequent inhibitory actions within the CNS (Kaada 1976). Evidence in favor of this view includes the observations that local anesthetics prevent the development of acupuncture analgesia (Chiang et al. 1973; Li 1973) and that a similar effect can be produced by peripheral nerve stimulation (Andersson and Holmgren 1975; Eriksson and Sjölund 1976). Furthermore, acupuncture of the legs of paraplegic patients does not cause analgesia in the upper body (Chang 1979; Peking Acupuncture Anaesthesia Co-ordinating Group 1973).

There is no obvious relationship between traditional acupuncture points and elements of the lymphatic or vascular system, but there is an association between the points and peripheral nerve branches (Fig. 12; Kaada 1976). A striking similarity has also been found between the locations of trigger points for myofascial pain and acupuncture points (Melzack et al. 1977). The long latency and duration of acupuncture analgesia suggests a humoral action, an idea that is supported by Chinese evidence that analgesia can be produced in animals that have not been subjected to acupuncture by intraventricular injection of cerebrospinal fluid (CSF) or by cross-circulation from animals given acupuncture (Kaada 1976; Han et al. 1980).

Acupuncture mechanisms are under active investigation in animal models (Cheng and Pomeranz 1980; Chiang et al. 1973; Dorman and Gage 1978; Fung et al. 1975; Galeano and Leung 1978; Galeano et al. 1979; Ha et al. 1981; Kaada 1974; Lee G.T.C. et al. 1976; McLennan et al. 1977; Peking Acupuncture Anaesthesia Co-ordinating Group 1973; Pomeranz and Chiu 1976; Pomeranz and Paley 1979; Toda et al. 1980), as well as in human subjects. Commonly used tests of the effect of acupuncture in animals are flexion reflexes, such as the jaw-opening reflex, and acupuncture points are selected by analogy with human points. In some experiments on animals, acupuncture seems to have no effect (Galeano and Leung 1978; Galeano et al. 1979; Kerr et al. 1978; Toda and Ichioka 1978), while in others there is a powerful action (e.g., Ha et al. 1981; Kerr et al. 1978; Toda and Ichioka 1978; Vierck et al. 1974). It is thought by at least one group that acupuncture results from activation mainly of muscle afferents rather than cutaneous afferents (Chiang et al. 1973). The analgesia is reported to be related to the activation of Aαβ or group II fibers (Toda and Ichioka 1978; Pomeranz and Paley 1979). The ascending pathway mediating acupuncture analgesia appears to ascend in the ventral part of the lateral funiculus contralateral to the side of the acupuncture point (Chiang et al. 1975). Toda et al. (1980) find that acupuncture in rats is more effective when loci comparable to those used in human acupuncture are needled, but that other sites produce some effects. Kerr et al. (1978) made a similar observation in cats, and in fact could get an equivalent effect from loci on or away from a traditional locus. It is also true in humans that

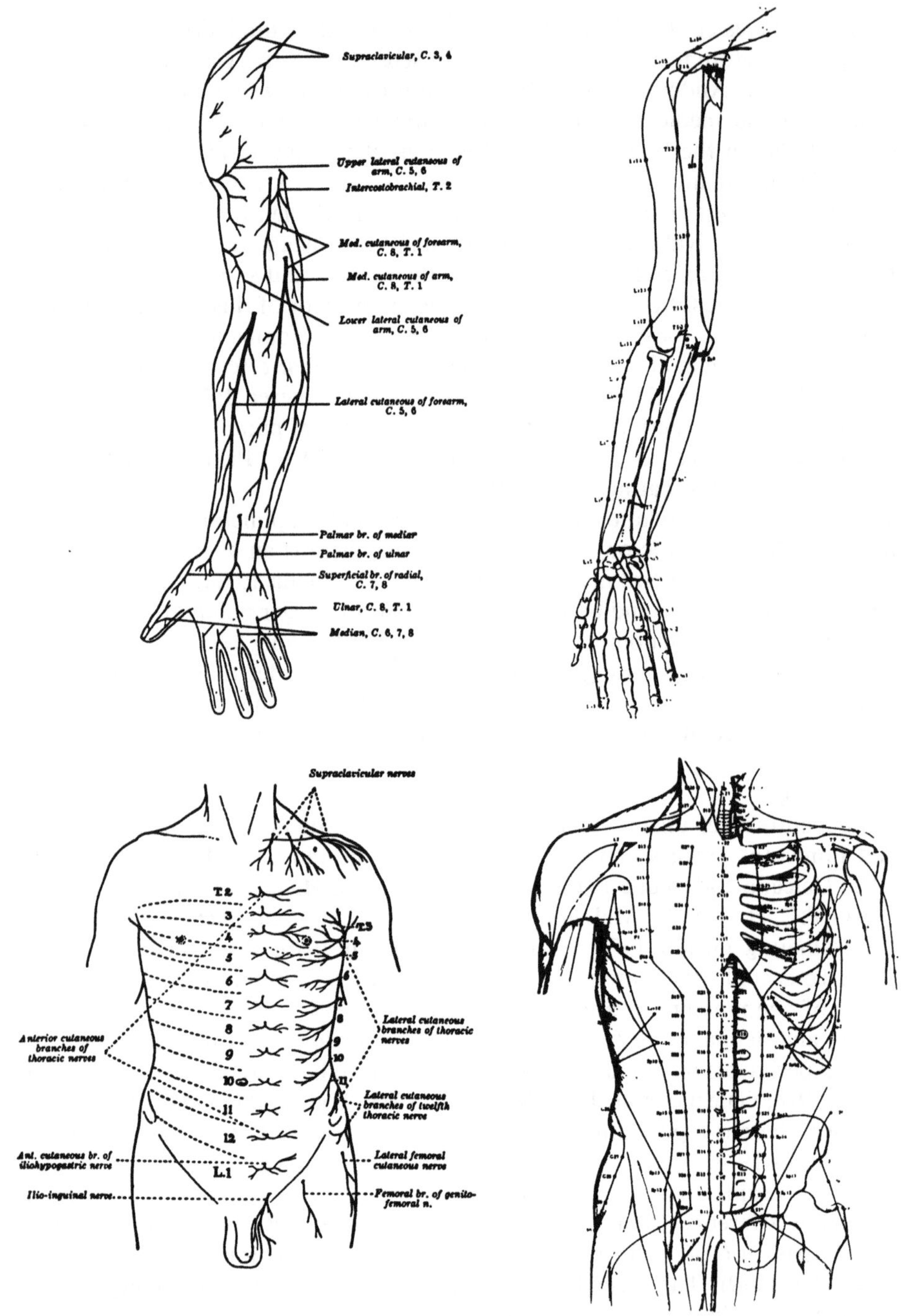

Fig. 12. Relationship between peripheral nerve branches (*left*) and acupuncture loci (*right*). (Kaada 1976)

acupuncture using points away from traditional loci can produce analgesia, although the analgesia is less (Croze et al. 1976).

Du and Chao (1976) distinguished between an inhibitory effect of electroacupuncture in cats on a viscerosomatic reflex during stimulation and a long-lasting inhibition following termination of stimulation (30 min after 5 min of electroacupuncture). Lesions were made in an effort to determine what CNS structures were required for each type of inhibition. Cerebellar removal generally had no effect. Intercollicular decerebration nearly eliminated the long-lasting inhibition in most animals, but had no effect on the inhibition produced during electroacupuncture stimulation. This result presumably explains the infrequent occurrence of acupuncture effects in the experiments of Pomeranz et al. (1977) and Kerr et al. (1978) in cats decerebrated at an intercollicular level. Du and Chao (1976) further showed that brain stem transection between the midbrain and pons abolished the after-inhibition, and either did not affect or somewhat reduced the inhibition during stimulation. The latter reflex inhibition was profoundly reduced by lesions transecting the lower medulla or upper cervical cord (although some inhibition remained even in spinalized animals; cf. Kerr et al. 1978; Pomeranz et al. 1977). Inhibition during electroacupuncture was also reduced markedly by bilateral sections of the lateral medulla passing through the region of the lateral reticular nucleus. Du and Chao (1976) suggest that this might be caused by the interruption of ascending tracts activated by acupuncture stimulation, although they could not rule out an effect on descending pathways. Midline lesions of the pons did not have an effect on either the after-inhibition or the inhibition during acupuncture. A midline lesion of the rostral and middle medulla eliminated the after-inhibition and severely depressed the inhibition during electroacupuncture. It was concluded that the after-inhibition of the reflex may require the participation of fore-brain structures in acupuncture analgesia in addition to structures in the medial part of the medulla. On the other hand, the inhibition occurring during acupuncture stimulation requires midline structures in the lower brain stem, such as the raphe nuclei and medial reticular formation.

A later study by Chiang et al. (1979) showed that lesions either of the NRM or of the raphe dorsalis or medianus (but not of the raphe pallidus or obscurus) produced a reduction in acupuncture analgesia lasting 1 – 2 weeks. The lesions were either electrolytic or chemical; in the latter case, the neurotoxin 5,6-dihydroxytryptamine was injected in order to restrict the lesion to serotonergic neurons.

The involvement of an opiate mechanism in acupuncture analgesia is strongly suggested by the fact that the analgesia can be antagonized in both humans and animals by naloxone (Fig. 13; Cheng and Pomeranz 1979; Fu et al. 1980; Ha et al. 1981; Han et al. 1980; Mayer et al. 1977; Pert et al. 1981; Pomeranz and Chiu 1976; Toda et al. 1980; Zhang et al. 1980; however, cf. Chapman et al. 1980; McLennan et al. 1977). Furthermore, endorphins are released during acupuncture (Pert et al. 1981; Zhang et al. 1980). Although some studies indicate that acupuncture produces a localized or at least regional analgesia based on the selection of a particular locus for needle insertion (Chang 1979; Chiang et al. 1973; Stewart et al. 1977), it has also been reported that the analgesia is widespread (Lynn and Perl 1977). This latter observation is consistent with the notion that the descending control systems may be responsible, since there is little evidence

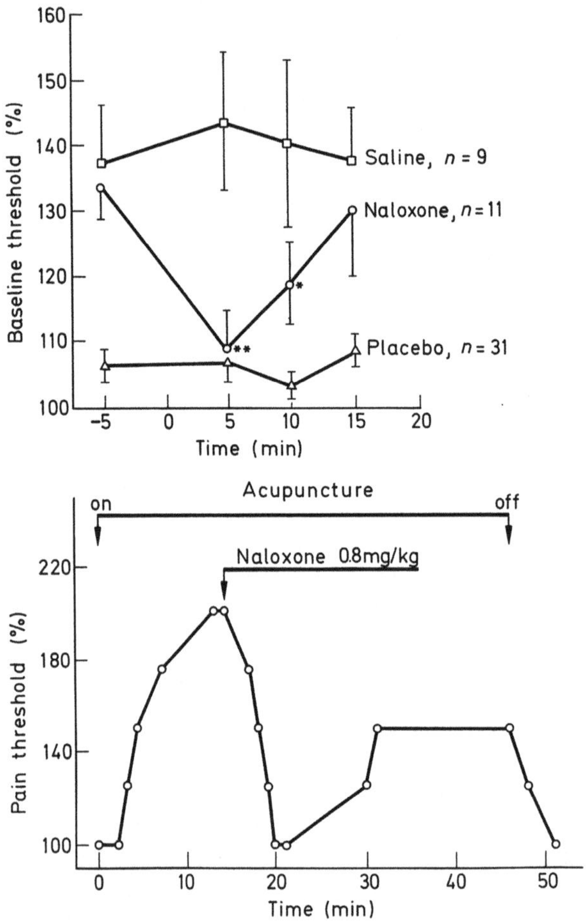

Fig. 13. Reversal of acupuncture analgesia by the opiate antagonist naloxone. Above, acupuncture was induced in human volunteers by inserting needles into the *Ho-ku* points (interosseus dorsalis I muscle between thumb and index finger) and rotating the needles intermittently for 30 min. Pain threshold to stimulation of tooth pulp was tested. While the pain threshold was elevated during acupuncture, either naloxone or saline was given; the naloxone reduced the pain threshold to the same level as seen in subjects not given acupuncture but only a placebo. (Mayer et al. 1977). *Below*, acupuncture was induced in awake, behaving macaque monkey by inserting and rotating a needle in the *Ho-ku* point. Analgesia was assessed using the jaw-opening reflex to tooth pulp stimulation. Naloxone administration caused a transient interruption of the acupuncture effect. (Ha et al. 1981)

for a precise somatotopographic organization of these pathways (Balagura and Ralph 1973; Dickenson et al. 1979; Hayes and Rustioni 1981; Hosobuchi et al. 1977a; Liebeskind et al. 1973; Lewis and Gebhart 1977a,b; Oliveras et al. 1974a; Richardson and Akil 1977a; however, cf. Mayer and Liebeskind 1974; Mayer et al. 1977; Watkins et al. 1980; Yaksh et al. 1976b).

Recently, it has been found that administration of D-amino acids, such as D-phenylalanine and D-leucine, can produce a naloxone-reversible analgesia

(Ehrenpreis et al. 1979) and can potentiate electroacupuncture (Cheng and Pomeranz 1980; Han et al. 1980). The D-amino acids may protect endorphins from enzymatic destruction.

It has been suggested that acupuncture involves the release of β-endorphin from the pituitary gland (Pomeranz et al. 1977). β-Endorphin injected into the peripheral circulation can exert an analgesic effect (Tseng et al. 1976). Furthermore, β-endorphin might gain access to the ventricular system by way of the median eminence (Bergland and Page 1978; Oliver et al. 1977). Alternatively, acupuncture may release not only pituitary β-endorphin but also CNS opioids (Pomeranz and Chiu 1976). In support of this is the finding that acupuncture analgesia persists in hypophysectomized animals (Fu et al. 1980).

Electrical stimulation of peripheral nerves has been used to produce analgesia in the West since classical times (Kane and Taub 1975; Kellaway 1946). Originally, the treatments involved the discharges of electric fish, such as the torpedo, the Nile catfish, or the electric eel.[1] After the invention of devices to produce electricity, it became fashionable in Europe and in the United States to use electricity in the therapy for various disorders, including pain (Fig. 14A). Electroanesthetic apparatuses were in common use during the nineteenth century, but fell into disrepute in the twentieth century (Kane and Taub 1975) until the report of Wall and Sweet (1967) reintroduced electrical stimulation of peripheral nerve as a valid therapeutic modality. Since then, transcutaneous nerve stimulation has become widely used (Fig. 14B; e.g., Augustinsson et al. 1977; Cauthen and Renner 1975; Cooperman et al. 1977; Ebersold et al. 1975; Eriksson and Sjölund 1976; Eriksson et al. 1979; Fields et al. 1974; Fox and Melzack 1976; Hiedl et al. 1979; Loeser et al. 1975; Long 1973; Long and Hagfors 1975; Mannheimer and Carlsson 1979; Melzack 1975b; Meyer and Fields 1972; Miles and Lipton 1978; Picaza et al. 1975; Rosenberg et al. 1978; Schuster and Infante 1980; Shealy 1974; Shealy and Maurer 1974; Sternbach et al. 1976; Strassburg et al. 1977; Sweet and Wepsic 1968; Thorsteinsson et al. 1977; Van der Ark and McGrath 1975; Woolf 1979). As in all forms of pain therapy, a reduction in pain due to the placebo effect must be kept in mind (Melzack 1975b; Thorsteinsson et al. 1978). Experimental studies show that electrical stimulation of peripheral nerves can produce analgesia in animal subjects, as well as in humans (Buckett 1979; Woolf et al. 1980).

It has been suggested by Andersson and Holmgren (1975) and by Sjölund and Eriksson (1976) that there are actually two forms of analgesia produced by transcutaneous nerve stimulation (cf. Cheng and Pomeranz 1979). When low-intensity, high-frequency electrical stimuli are used, an analgesia is produced that is restricted to the same body region as that stimulated. This analgesia is not reversed by naloxone and is detectable soon after stimulation is begun (Andersson and Holmgren 1975; Eriksson and Sjölund 1976; Pertovaara and Kemppainen 1981; Shealy and Maurer 1974; however, cf. Buckett 1979). There are generally paresthesias referred to the area of pain relief. This is the type of analgesia de-

1 According to Kellaway (1946), Scribonius Largus suggested that "For any type of gout a live black torpedo should, when pain begins, be placed under the feet. The patient must stand on a moist shore washed by the sea and he should stay like this until his whole foot and leg up to the knee is numb. This takes away present pain and prevents pain from coming on if it has not already arisen"

a

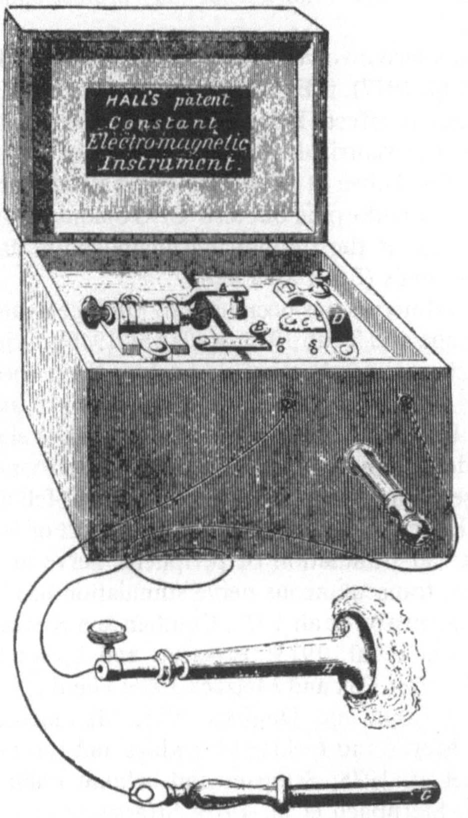

Fig. 14a. Legend see p. 33

scribed by Wall and Sweet (1967), and a similar effect is produced by electrical stimulation of the dorsal columns (Fox 1974; Lindblom and Meyerson 1975; Long and Hagfors 1975; Nashold and Friedman 1972; Nielson et al. 1975a,b; Shealy et al. 1970; Sweet and Wepsic 1974), the ventral posterior nuclei of the thalamus (Boëthius et al. 1976; Hosobuchi et al. 1973; Mazars 1975), or the internal capsule (Adams et al. 1974; Fields and Adams 1974). A peripheral mechanism has been proposed for the action of transcutaneous nerve stimulation (Campbell and Taub 1973; Ignelzi and Nyquist 1976; Taub and Campbell 1974; Torebjörk and Hallin 1974). However, Janko and Trontelj (1980) provide evidence that changes in compound action potentials during transcutaneous nerve stimulation result from desynchronization and not from failure of conduction. Furthermore, the similar effects of central stimulation suggest a central mechanism.

On the other hand, high-intensity, low-frequency transcutaneous nerve stimulation produces a more global analgesia that is naloxone-reversible (Sjölund and Eriksson 1976, 1979). This form of analgesia is only detectable after stimulation for some 20 min, and it may last for hours to days after stimulation is discontinued. This type of analgesia resembles acupuncture analgesia and presumably results from the same mechanism (Andersson and Holmgren 1975; Chap-

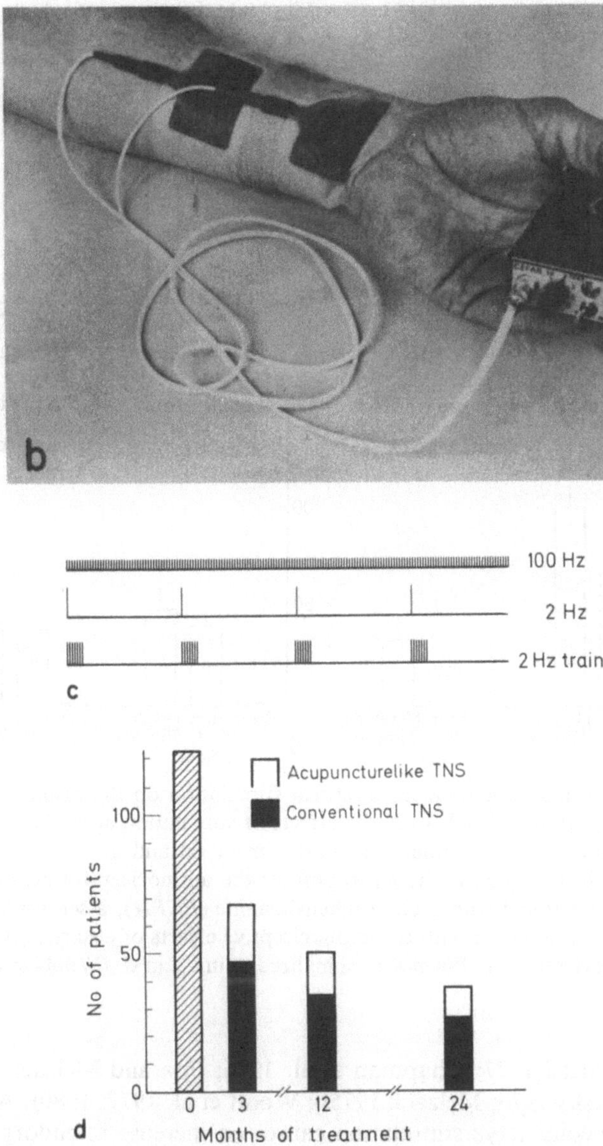

Fig. 14. a Garratt's electroanesthesia machine, 1860. (Kane and Taub 1975) **b** Modern transcutaneous nerve stimulator. **c** Pulse train configurations: *above* for high-frequency transcutaneous nerve stimulation (*TNS*); *lower two* for acupuncture-like TNS. **d** Number of patients with chronic pain starting treatment with TNS (*hatched*) and continuing. (Eriksson et al. 1979)

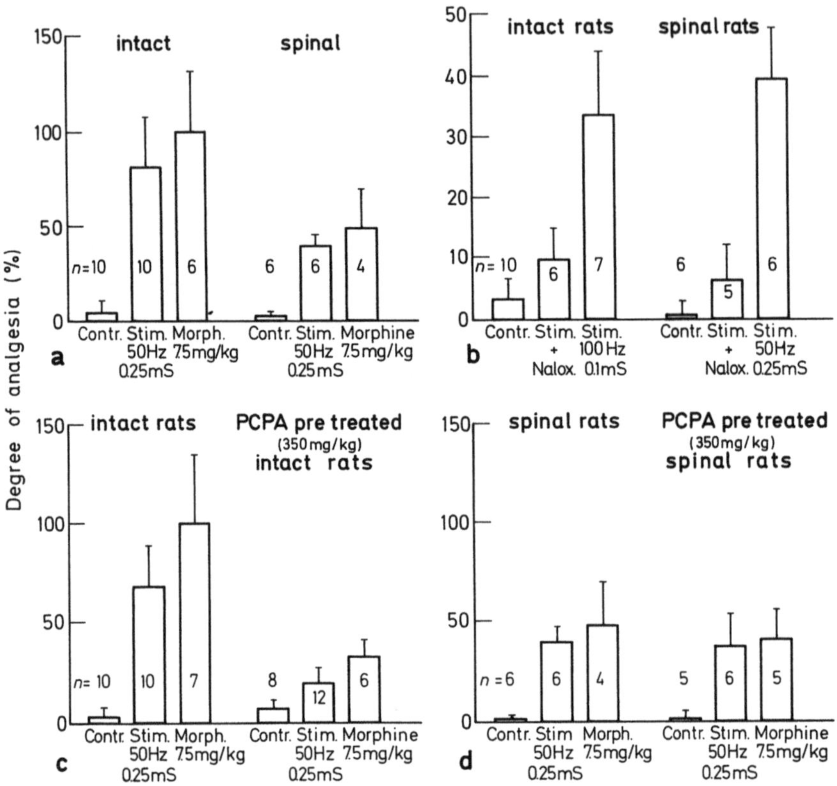

Fig. 15a–d. Effect of percutaneous electrical stimulation on the response in the rat to noxious heat applied to the tail. In **a**, the effects of stimulation at the base of the tail are compared with those of morphine, both in the intact rat and in rats having spinal cord transections. In **b**, naloxone is shown to prevent the antinociceptive action of electrical stimulation. Pretreatment with *p*-chlorophenylalanine (*PCPA*), a serotonin synthesis inhibitor, is shown to interfere with the antinociceptive effects of electrical stimulation and morphine in intact rats in **c**, but not in spinalized animals in **d**. (Woolf et al. 1980)

man and Benedetti 1977; Chapman et al. 1976; Fox and Melzack 1976; Goldberger and Tursky 1976; Melzack 1975b; Woolf et al. 1977, 1980). Acupuncture-like transcutaneous nerve stimulation causes an increase in endorphin levels in the CSF (Sjölund et al. 1977), suggesting that pain relief may be due at least in part to activation of an opiate mechanism. It has been reported that transcutaneous nerve stimulation causes an elevation of CSF β-endorphin levels even in the absence of pain (Salar et al. 1981).

In a study on rats, Woolf et al. (1980) were able to produce an acupuncture-like analgesia using either low-frequency, high-intensity or high-frequency, lower-intensity electrical stimulation. The animals were tested for response to noxious heat by dipping the tail into water at 49 °C. The response in intact animals included both a tail flick, presumably a reflex, and a later strong tail flexion, accompanied by struggling and vocalization. When the same test was done in spinalized rats, there was only a strong flexion of the tail that occurred at a shorter latency than in intact rats. Morphine increased the latencies of the responses, both in the intact and in the spinalized animals, although there was a smaller effect in the spinalized rats (Fig. 15A). Percutaneous electrical stimulation at the base of the tail (50 Hz, 0.25 ms pulses, 10 – 15 V) produced almost as great an increase in response latency as did morphine, both in intact and in spinalized animals (Fig. 15A). Low-frequency stimulation had less effect even with stronger stimuli. The analgesia was detectable after 15 min stimulation and was greater at 30 – 45 min. Following stimulation, the analgesia was still maximal at 15 min, less at 30 min, and gone at 40 min. Pretreatment with naloxone prevented the analgesia due to electrical stimulation, both in intact and in spinalized animals (Fig. 15B). Pretreatment with *p*-chlorophenylalanine also prevented the analgesia in intact animals, but not in spinalized animals (Fig. 15C,D; see Section 3). It was concluded that peripheral stimulation can activate antinociceptive mechanisms at both supraspinal and spinal levels, and that these mechanisms depend upon opiate release at both levels, but that serotonin is involved just in the supraspinal mechanism.

In addition to the analgesia produced by electrical stimulation of peripheral afferent nerve fibers, it is possible to produce analgesia using more natural forms of stimulation. One such approach is the use of vaginal probing in female rats to induce analgesia (Komisaruk et al. 1976; Komisaruk and Wallman 1977). This analgesia is enhanced following administration of estradiol (Crowley et al. 1976). It is likely that this form of analgesia plays a role in mating behavior and in the birth process (Crowley et al. 1976; Komisaruk and Wallman 1977). Although an earlier report suggested that this kind of analgesia is not antagonized by naloxone (Crowely et al. 1977a), a recent paper indicates that it is (Hill and Ayliffe 1981). Thus vaginal probing may prove to be a means of activating the endogenous opiate control systems. Komisaruk and Wallman (1977) note the resemblance between this and acupuncture analgesia.

2.5 Stress-Induced Analgesia

It is common experience that individuals may experience injuries that are not initially painful if sustained during a stressful event. This is true for battlefield wounds (e.g., Beecher 1959), sporting accidents, car wrecks, and other traumas. Stress-induced analgesia is being studied in animal models by a number of investigators (Amir and Amit 1979; Baizman et al. 1979; Bodnar et al. 1978a,b). Stress is induced by immobilization, inescapable foot shock, cold water swims, food deprivation, hypoglycemia (induced by insulin or 2-deoxy-D-glucose), intraperitoneal injections of hypertonic saline, or rotation. Problems in such studies are that the various means for inducing stress are not equivalent and that there is a distinct possibility of interaction between the means used to induce stress and the effect of peripheral stimulation per se in producing analgesia. Evidence that the effects of stress vary with the means used to induce stress includes the observation that opiate systems may or may not be involved, depending upon the conditions of the experiment (Amir and Amit 1978; Bodnar et al. 1978b,c,d, 1979b; Chance and Rosecrans 1979; Hayes et al. 1978; Lal et al. 1978; Lewis et al. 1980; Madden et al. 1977; Spiaggia et al. 1979; Spiler and

Molitch 1980). Serotonin and prostaglandins have been implicated in immobilization analgesia (Bhattacharya et al. 1978). Endocrine changes are likely to be involved, as well as neural events (Bernston and Berson 1980; Bodnar et al. 1979a,c, 1980; Rossier et al. 1977, 1979). One mechanism proposed is release of substances from the pituitary gland (Amir and Amit 1979; Bodnar et al. 1979a,c; Millan et al. 1980). At least some forms of stress-induced analgesia do not require integrity of pathways descending from the brain stem in the dorsolateral funiculi (Hayes et al. 1978).

2.6 Hypnotic Analgesia

It is possible in susceptible individuals to induce analgesia by hypnosis. Such an approach has been useful clinically to provide pain relief in such circumstances as childbirth, dental procedures, and burns; one distinct advantage of hypnosis is the avoidance of anesthetic drugs that might constitute an unacceptable risk under certain circumstances (Crasilneck et al. 1955; Finer 1974; Finer and Nylén 1961; Hilgard 1975; Orne 1974).

Hypnotic analgesia has been reported not to involve an opiate mechanism, since it is not naloxone-reversible (Barber and Mayer 1977; Goldstein and Hilgard 1975; however, cf. Stephenson 1978). Furthermore, according to Hilgard (1975; Hilgard and Hilgard 1975), subjects of hypnotic analgesia are able to experience nociceptive input during hypnosis, and so the analgesia may be based on alterations in perceptual processing at higher levels of the nervous system. Thus the mechanisms of hypnosis falls outside the scope of this review.

2.7 Conclusions

1. In order to investigate how pain can be controlled, it is necessary to devise means for measuring pain. This is difficult in humans and even more difficult in animals. In addition to methodological problems, there are also ethical issues.
2. The term "analgesia" means the absence of pain. This term is convenient for describing the outcome of experiments on animal subjects, but there are objections to its use for this purpose.
3. Commonly employed tests of "analgesia" in animals include several forms of the flexor withdrawal reflex. Such tests include the pinch test, tail flick, back skin twitch, jaw-opening reflex, and writhing response.
4. Other tests of "analgesia" in animals involve not only reflex responses but also some degree of higher-order processing. These tests include the hot plate test, the vocalization test, shock titration, and many others.
5. In human studies, the level of pain in patients can be quantitated using systematic verbal reports or analogue scales. In judging the effectiveness of treatments meant to reduce pain, the placebo effect must be taken into account.
6. Experimental pain can be induced in humans using a variety of noxious stimuli. The effects of painful stimuli and of analgesic agents can be measured

in the same ways as have been used for clinical pain. Alternatively, reflex responses can be recorded. However, the flexion reflex in humans is not invariably correlated with subjective reports of pain, since descending modulatory systems or even opiate drugs can alter pain or the flexion reflex differently. Another approach to the study of experimental pain in humans is the application of signal detection theory. The interpretation of such studies is controversial.

7. Animal experiments can be helpful in the study of neural correlates of pain transmission and analgesia. Such work can be done in awake, behaving animals or in reduced, often anesthetized animals. Each of these approaches has limitations and advantages.
8. Stimulation within certain regions of the brain can produce analgesia in animals and in human patients. This phenomenon is called "stimulation-produced analgesia" (SPA).
9. Brain sites that can be associated with SPA include regions near the midline, such as the periventricular and periaqueductal gray and the raphe nuclei. In addition, SPA results from stimulation in parts of the reticular formation and in a variety of other brain regions.
10. One difficulty in localizing the neural pathways responsible for SPA is the problem of what elements are being stimulated by electric shocks delivered through an electrode placed in a particular site. Both neuron cell bodies in a given nucleus and axons of passage can be activated. Furthermore, stimulus spread is inevitable when stimulus currents are large. Some of these problems can be alleviated by use of chemical stimuli.
11. Some of the characteristics of SPA are that:
    a) Analgesia may outlast stimulation, depending on stimulus duration and intensity.
    b) Onset of analgesia may be gradual or immediate.
    c) Analgesia can occur in the absence of other sensory changes, or alterations in motor activity, emotion, attention, or motivation.
    d) There is probably not a somatotopic relationship between stimulus site and the distribution of analgesia, although this is controversial.
12. Lesion studies do not always produce results that can be predicted from studies of SPA. For example, lesions of the periaqueductal gray (PAG) do not result in hyperalgesia, but rather hypalgesia or no change. The PAG is involved not only in SPA but also in aversive reactions, and so lesions may produce cancellation of opposing actions. Alternatively, the PAG might not be tonically active and so a lesion might have no overt effect.
13. Stimulation-produced analgesia can be produced in humans. There seem to be different results, depending upon whether SPA is the result of stimulation in the periventricular gray (PVG) or in the ventrobasal complex of the thalamus or internal capsule. For example, stimulation of the PVG does not result in paresthesias, whereas stimulation in the ventral thalamus or internal capsule does.
14. Analgesia has been produced in clinical pain states for centuries by administration of opiate drugs. Only recently has evidence been obtained that opiates act on opiate receptors in nervous tissue.

15. The presence of opiate receptors in nervous tissue appears related to the fact that there are also endogenous opioid substances in nervous tissue. These opioid substances include methionine- and leucine-enkephalin, β-endorphin, dynorphin, and Kyotorphin. These and perhaps other opioid peptides are candidate neurotransmitters or modulators.
16. Opiate receptors and endogenous opioid peptides are found in regions of the CNS that are involved in SPA or in nociceptive processing. It has been proposed that opiate mechanisms may be involved in pain mechanisms. However, opiate systems are likely to have many other functions, and neural systems not utilizing opiate mechanisms are also likely to be involved in the modulation of pain.
17. Analgesia can be produced not only by brain stimulation or opiate administration, but also by peripheral stimulation. Several techniques have been used to activate afferent nerve fibers in the periphery that cause analgesia. These include acupuncture, transcutaneous nerve stimulation, and vaginal probing.
18. Acupuncture means puncturing the skin with one or more needles. It is an ancient Chinese technique for treatment of various pathological conditions, including pain. However, analgesia results not from puncturing the skin per se, but from manipulation of the needle or passing a current through it.
19. The traditional Chinese explanation of acupuncture analgesia involves the redistribution of energy along channels or meridians that are thought to be in longitudinal array below the body surface. However, there are no anatomical structures corresponding to the channels. On the other hand, the best loci for acupuncture are near peripheral nerve branches. Since acupuncture is prevented by local anesthesia, and since a similar effect can be produced by peripheral nerve stimulation, acupuncture analgesia is certainly due to central effects of afferent nerve fiber volleys.
20. Acupuncture mechanisms are under active investigation in animal models. The afferent fiber types and central pathways involved are being studied. Brain structures implicated include sites thought to participate in SPA and opiate analgesia.
21. The fact that acupuncture can be counteracted in humans and in animals by naloxone, an opiate antagonist, suggests that an opiate mechanism may be involved in acupuncture analgesia. Substances that prevent enzymatic degradation of endorphins enhance acupuncture analgesia.
22. A part of the acupuncture analgesia may involve the release of β-endorphin from the pituitary gland.
23. Transcutaneous nerve stimulation involves electrical stimulation of afferent nerve fibers. This method has been employed in the West since classical times for pain relief.
24. There appear to be two forms of transcutaneous nerve stimulation. One typically involves high-frequency, low-intensity stimulation over a peripheral nerve that distributes to a painful region. This type of stimulation produces a localized analgesia that is rapid in onset, is not reversed by naloxone, and is associated with paresthesias. The other type involves low-frequency, high-intensity stimulation. It produces a more global analgesia that is naloxone-reversible; rather than producing paresthesias, this type of electrical stimula-

tion is often somewhat painful. Endorphins are found in elevated amounts in the cerebrospinal fluid after this type of stimulation.

25. It has been suggested that the naloxone-reversible form of transcutaneous nerve stimulation is a form of acupuncture.
26. Analgesia due to acupuncture-like transcutaneous nerve stimulation in intact animal subjects is prevented by either naloxone or by blockade of serotonin synthesis. A similar "analgesic" effect is seen in animals after spinal cord transection, and this is also blocked by naloxone, but not by interference with synthesis of serotonin.
27. "Analgesia" can be produced in female rats by vaginal probing. This form of analgesia may be naloxone-reversible. The analgesia resulting from vaginal probing may be of functional value in reproductive behavior, including parturition.
28. Conditions of stress can also lead to analgesia. This is true in humans and in animal subjects. There are probably several mechanisms of stress-induced analgesia, some involving opiate mechanisms and some not. Hormonal changes mediated through the endocrine system may be important, in addition to neural mechanisms.
29. Hypnotic analgesia can be very useful in certain clinical situations. This form of analgesia may not involve an opiate mechanism and may be due to changes in perceptual processing rather than to the action of descending control systems.

3 Pharmacology of Analgesia Due to Descending Control Systems

3.1 Overview

Prior to 1964, there was little reason to doubt that opiate analgesics, such as morphine, had a rather generalized action on the CNS. Furthermore, the fact that electrical stimulation of a focal region of the brain could prevent pain was not known until 1969. Opiate receptors were first convincingly demonstrated in brain tissue in 1973, and endogenous opioid peptides were first isolated from brain in 1975. With such a short time-frame, it should not be surprising that our understanding of the details of the organization of the neural systems responsible for producing the various forms of analgesia described in Section 2 is still very fragmentary, and that there are many apparently conflicting pieces of evidence in the literature on this subject. This problem is particularly notable with respect to the pharmacological evidence considered in this section.

The first topic to be considered is the possible relationship between the neural mechanism responsible for SPA and that responsible for opiate analgesia. After this, evidence will be reviewed concerning the view that SPA can involve monoaminergic synaptic transmission. Next, the possible role of the monoamines in opiate analgesia will be discussed. Then, hints that other synaptic transmitters or "modulators" may be involved in the analgesia systems will be noted in a somewhat anecdotal fashion. Finally, the pharmacology of analgesia due to acupuncture and transcutaneous nerve stimulation will be considered briefly.

3.2 Relationship Between SPA and Opiate Analgesia

Tsou and Jang (1964) were the first to show that morphine can cause analgesia when injected in small amounts directly into the brain. Studies employing intraventricular administration of morphine in the laboratory of Herz (e.g., Herz et al. 1970; Herz and Teschemacher 1971; Teschemacher et al. 1973; Vigouret et al. 1973) showed that the most sensitive sites for morphine action are adjacent to the third and fourth ventricles and the cerebral aqueduct, being especially concentrated around the rostral fourth ventricle (Fig. 16). For part of this work, the nociceptive response tested was the licking response of rabbits to electrical stimulation of tooth pulp. Thus a possible site of action of the opiates was directly on the trigeminal system in these experiments (Herz and Teschemacher 1971). However, similar results were obtained when the nociceptive response tested was the hind limb flexion reflex (Teschemacher et al. 1973; Vigouret et al. 1973).

Using the microinjection technique, Yaksh and his colleagues further delimited the distribution of active sites for opiate action (Pert and Yaksh 1974, 1975; Yaksh and Rudy 1978; Yaksh et al. 1976b; Yeung et al. 1977). These were chiefly in the PAG and PVG (Fig. 17). The threshold dose for a morphine action was in the order of μg (Yaksh et al. 1976b). The possibility that the injected drug actually had its effect at a distance from the injection site after escape into the

Fig. 16. Distribution of ^{14}C-morphine as determined by radioautography. In *column I*, the opiate was injected intraventricularly, and it produced a strong analgesic effect as judged by changes in the hind limb flexion reflex in the rabbit. In *column II*, the opiate was injected intracisternally and there was no analgesia. *Column III* shows the difference in labeling (area in II subtracted from area in I), indicating the most likely sites of action of morphine. * indicates the caudal end of the cerebral aqueduct. (Teschemacher et al. 1973)

ventricular system (Herz and Teschemacher 1971) was rejected on the grounds that negative results were seen after microinjections placed at similar distances from the ventricular system as for sites giving positive results (Yaksh et al. 1976b). Sometimes electrical stimulation of the same sites that caused analgesia with microinjection of morphine could be shown to result in SPA (Yeung et al. 1977; see also Mohrland and Gebhart 1980). However, more often only one manipulation was effective (Yeung et al. 1977), indicating at least a partial physical separation of the mechanisms underlying SPA and opiate analgesia within the PAG. Similar observations were made by Lewis and Gebhart (1977a,b). These workers suggest that SPA may result from stimulus escape to fiber systems adjacent to the PAG (Lewis and Gebhart 1977b), and they reraise the question of

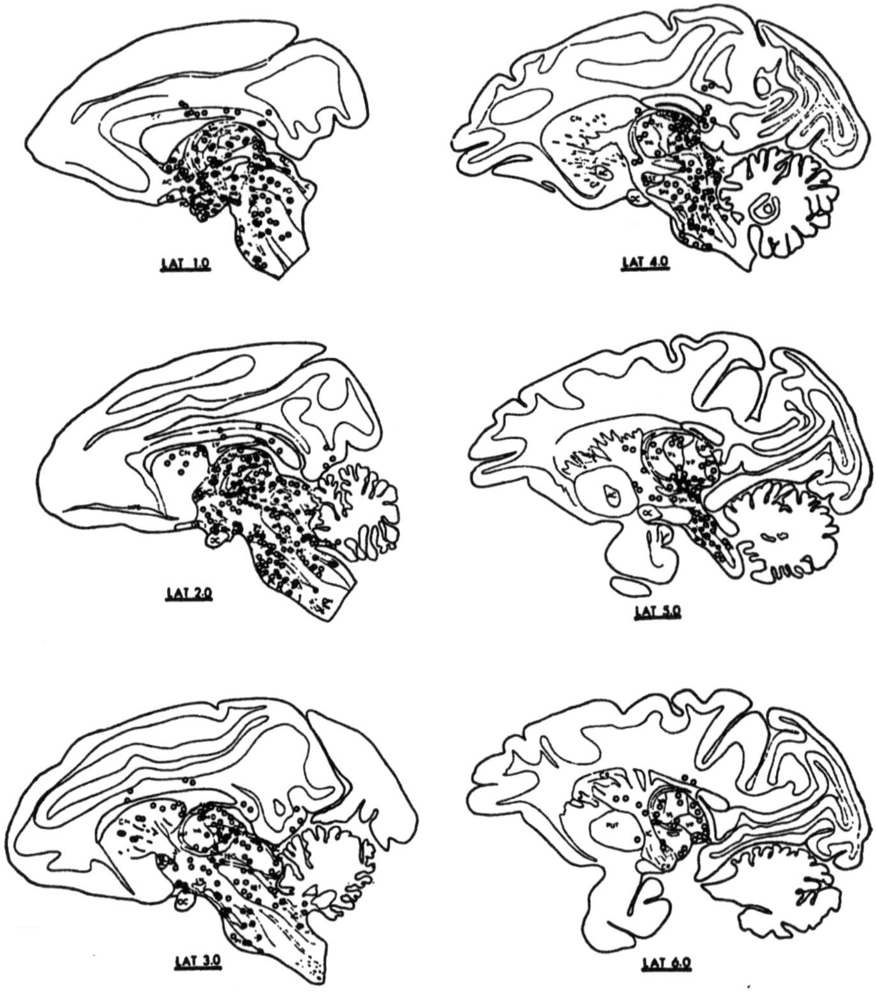

Fig. 17. Brain sites tested for the effects of morphine microinjections (40 μg), using the shock titration method for demonstrating analgesia in the monkey. No effect is indicated by *open circle*; moderate increase in nociceptive threshold by *filled circle*; maximum effect by *asterisk*. (Pert and Yaksh 1974)

leakage of morphine into the ventricular system following microinjection. A number of groups have reported analgesia following microinjections of morphine into the PAG or PVG; sometimes the animals are hyperreactive to other stimuli but they nevertheless show a reduced response to noxious stimuli (Criswell 1976; Herz et al. 1970; Jacquet and Lajtha 1973, 1974; Lewis and Gebhart 1977a,b; Sharpe et al. 1974; Yaksh et al. 1976b). Hyperreactivity has also been found during SPA (e.g., Oliveras et al. 1979). The shortest latency and most potent analgesic actions of morphine injected into the PAG were associated with sites in the ventrolateral part of the caudal PAG (Yaksh et al. 1976b). Injections in this region tended to produce a whole-body analgesia, whereas injections in the rostral PAG resulted in an analgesia over just the dorsal aspect of the face and upper extremities.

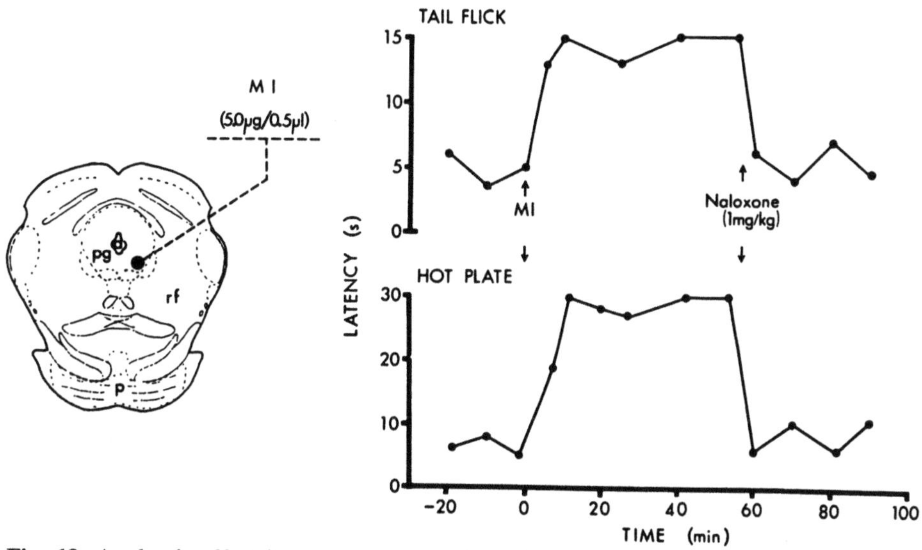

Fig. 18. Analgesic effect in a rat of a microinjection of morphine (5 μg) into the periaqueductal gray. The tests used are the tail flick (*above*) and the hot plate (*below*). The time of morphine injection is indicated by *MI* and an *arrow*. The analgesic action is reversed by naloxone, which was given systematically at the time of the *second arrow*. (Yaksh et al. 1976b). *p*, pons; *pg*, periaqueductal gray; *rf*, reticular formation

A key test whether the opiates or endogenous opioid substances have a specific action on opiate receptors is their susceptibility to antagonism by opiate blockers (Hayes et al. 1977b; Markowitz et al. 1976; Martin 1967; Takemori 1974). Some of these substances, including naloxone and naltrexone, do not seem to have an overt action themselves on pain in the absence of an opiate action (Goldstein et al. 1976; Grevert and Goldstein 1977a, 1978; Gritz et al. 1976; however, cf. Frederickson et al. 1977; Jacob et al. 1974; Walker et al. 1977), suggesting that the endogenous opiate mechanisms are not usually tonically active and are not easily engaged in experimental pain tests. The analgesic effect of morphine when microinjected into the PAG is antagonized by naloxone, indicating that the action of morphine is due to an opiate-receptor-mediated change (Fig. 18; Yaksh et al. 1976b). Behbehani and Fields (1979) were able to produce analgesia in rats by injecting glutamate into the PAG. This presumably avoided the problem of stimulation of axons of passage, one of the hazards of electrical stimulation. Furthermore, the analgesia was prevented by naloxone (after an unexplained delay), suggesting that the glutamate injections engaged an opiate mechanism. Most of the early microinjection studies explored the rhombencephalon in only a cursory fashion, and so it is not surprising that opiate-sensitive regions in this region have since been emphasized (Fig. 19), including areas in the reticular formation (Akaike et al. 1978; Takagi et al. 1977, 1978) and the NRM (Dickenson et al. 1979; LeBars et al. 1980; however, cf. Akaike et al. 1978). The region of the reticular formation ventral to the nucleus gigantocellularis proper seems to be the most sensitive place in the brain to microinjections of morphine, with analgesia resulting from as little as 2-ng doses (Akaike et al. 1978). This region is

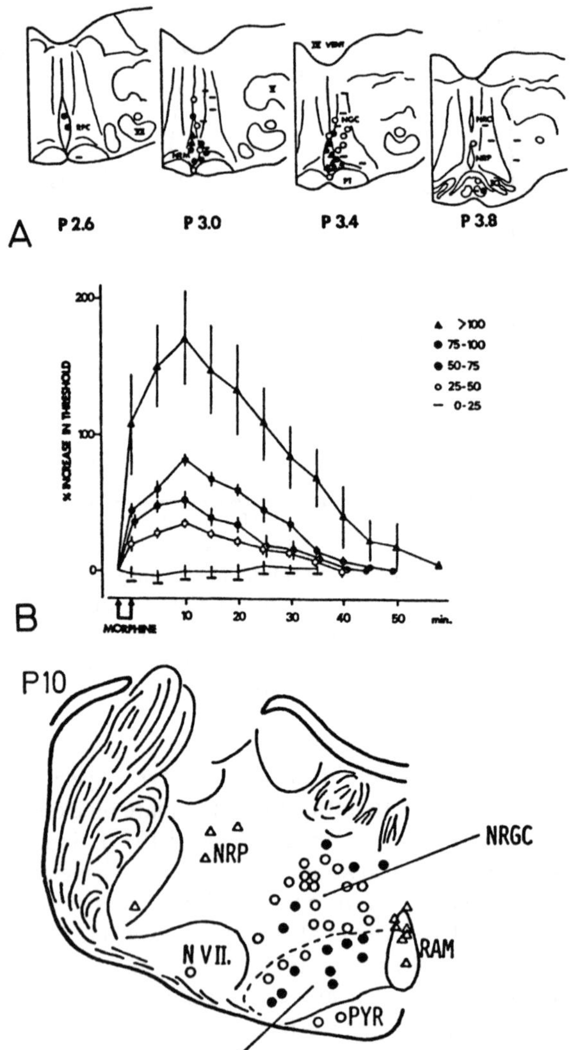

Fig. 19. The sites at which microinjections of morphine (5 μg) produced analgesia in rats are shown in **A**. The degree of analgesia is expressed as a percentage of that caused by systemic injections of morphine (3 mg/kg). Analgesia was judged by the vocalization and the pinch tests. The time course of action of microinjected morphine is shown for different groups of animals in **B**. (Dickenson et al. 1979). **C** Shows the sites in the rat brain stem reticular formation that did (*filled symbols*) or did not (*open symbols*) cause analgesia following microinjections of morphine (15 ng, *circles*; 380 ng, *triangles*). Analgesia was tested by tail pinch. (Akaike et al. 1978). *NRM, RAM*, nucleus raphe magnus; *NGC* and *NRGC*, n. ret. gigantocellularis; *NRPG*, n. ret. paragigantocellularis, *PYR*, pyramid; *NVII*, facial nucleus

referred to in the rat as the nucleus paragigantocellularis lateralis, and it appears to be equivalent to the magnocellular part of the nucleus gigantocellularis of the cat (Berman 1968). The actions of opiates injected into the NRM or the reticular formation are prevented or reversed by naloxone (Akaike et al. 1978; Dickenson et al. 1979; Takagi et al. 1977, 1978).

It is thought that most, if not all, of the action of low doses of systemically administered morphine is mediated by an action on the opiate receptors in the brain. Evidence for this includes the observations that spinal cord potentials evoked by noxious stimuli and the tail flick response are reduced by lower doses of morphine if the spinal cord is intact than if the cord is transected (Dewey et al. 1969; Irwin et al. 1951; Proudfit and Levy 1978; Satoh and Takagi 1971; Takagi

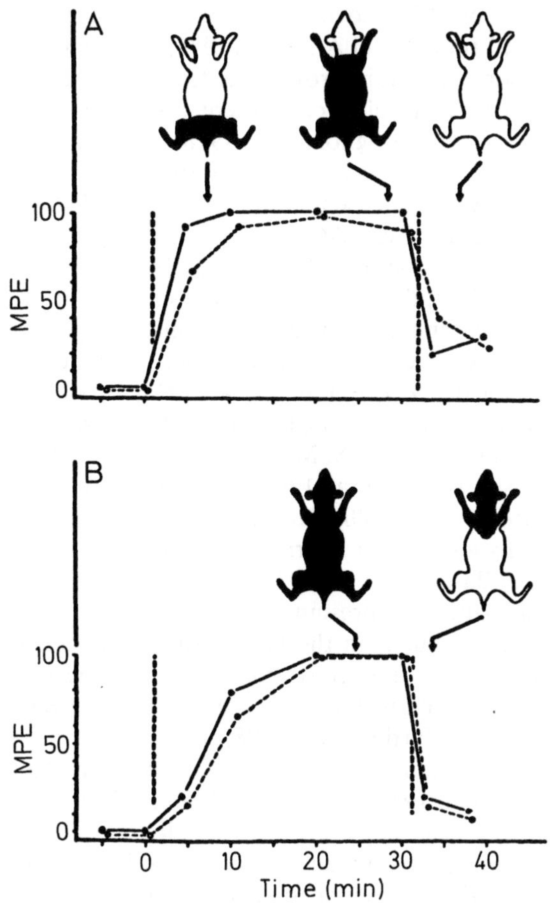

Fig. 20. Effect of intrathecally injected morphine (15 µg) is shown in **A** and that of systemically administered morphine (20 mg/kg) in **B**. Analgesia was measured by the tail flick (*solid lines*) and hot plate (*dashed lines*) tests. The distribution of analgesia is shown by the *blackened areas* on the figurines as revealed by the pinch test. Morphine was given at the times indicated by the *first vertical dashed lines,* and naloxone was given at the *second lines.* For the animal in **A**, naloxone was given systemically (1 mg/kg), while for the experiment in **B** it was given intrathecally (5 µg). (Yaksh and Rudy 1977)

et al. 1955), and that lesions of the PAG or of the raphe nuclei reduce or eliminate the analgesia produced by systemic morphine injections (Adler et al. 1975; Dostrovsky and Deakin 1977; Garau et al. 1975; Proudfit and Anderson 1975; Samanin and Bernasconi 1972; Samanin et al. 1970; Yaksh et al. 1977a; however, cf. Bläsig et al. 1973; Proudfit 1980). There is cross-tolerance between intravenously and intracerebrally injected morphine (Jacquet and Lajtha, 1976). Furthermore, injections of opiate antagonists into the PAG or NRM antagonize the analgesic effects of systemic injections of morphine (Dickenson et al. 1979). With large doses of morphine, actions on the spinal cord become prominent (Barton et al. 1980; Irwin et al. 1951; Proudfit and Levy 1978; Satoh and Takagi 1971). A selective action of morphine on the spinal cord can be obtained by administration of morphine, met-enkephalin, or β-endorphine intrathecally (Fig. 20; Tang and Schoenfeld 1978; Wang et al. 1979; Yaksh 1978; Yaksh and Henry 1978; Yaksh et al. 1977b; Yaksh and Rudy 1976, 1977).

The pathways for SPA and for opiate analgesia originating from the midbrain level appear to include a synaptic link in the medulla, probably in the NRM and perhaps the adjacent reticular formation (see Sections 4 and 5; Anderson SD et al. 1977; Behbehani and Fields 1979; Behbehani and Pomeroy 1978; Fields and

Anderson 1978; Lovick et al. 1978; Oleson et al. 1978; Pomeroy and Behbehani 1979; Proudfit and Anderson 1975; Shah and Dostrovsky 1980). Although there are some midbrain projections from the PAG and adjacent reticular formation to the cervical spinal cord, there are few direct projections to lumbosacral levels of the cord (Castiglioni et al. 1978; Crutcher et al. 1978; Edwards 1975; Kneisley et al. 1978; Kuypers and Maisky 1975; Martin et al. 1979). Presumably, descending activity from the midbrain to the spinal cord needs to be relayed in the lower brain stem. The PAG is known to project to the NRM and to the adjacent reticular formation (Abols and Basbaum 1981; Gallager and Pert 1978; Pierce et al. 1976; Ruda 1975). There are spinal projections from several nuclei of the rhombencephalon, including the nuclei raphe magnus, obscurus, and pallidus, and several regions of the reticular formation (Basbaum et al. 1978; Basbaum and Fields 1979; Bobillier et al. 1976; Brodal et al. 1960; Crutcher et al. 1978; Goode et al. 1980; Hayes and Rustioni 1981; Kuypers and Maisky 1975, 1977; Leichnetz et al. 1978; Martin G. F. et al. 1979; Martin R. F. et al. 1978; Nyberg-Hansen 1965; Petras 1967; Tohyama et al. 1979a,b; Torvik and Brodal 1957; Watkins et al. 1980). However, both SPA and opiate analgesia are reduced following lesions that interrupt the dorsal parts of the lateral funiculi (Barton et al. 1980; Basbaum et al. 1976, 1977; Hayes et al. 1978). Thus the descending analgesia pathway is presumed to descend in this part of the spinal cord white matter. The main nuclei in the lower brain stem that project to the cord through the dorsolateral funiculi are the NRM and the magnocellular part of the nucleus gigantocellularis (Basbaum et al. 1976, 1978; Basbaum and Fields 1979; Goode et al. 1980; Martin et al. 1978; Tohyama et al. 1979a,b; Watkins et al. 1980). Synaptic endings of the descending axons are found in laminae I, II, and V (Basbaum et al. 1978; Goode et al. 1980). It is the raphe-spinal and "dorsal" reticulospinal tracts that are now generally considered to be the main candidates to convey the descending volleys responsible for SPA.

So far, the evidence cited only indicates that SPA and opiate analgesia depend on similarly arranged descending pathways. However, there is reason to believe that these two methods of causing analgesia may share a common neural mechanism (Mayer 1975). For instance, SPA can be reduced by the opiate antagonist, naloxone (Adams 1976; Akil et al. 1976, 1978a; Hosobuchi et al. 1977a; Oliveras et al. 1977b; Pert and Walter 1976; Zorman et al. 1981; however, cf. Gebhart and Toleikis 1978; Hayes et al. 1979a; Yaksh et al. 1976c). Furthermore, under some circumstances, there is likely to be a tonic release of endogenous opioids, since naloxone administration can result in hyperalgesia (Frederickson et al. 1977; Grevert and Goldstein 1977b; Jacob et al. 1974; Walker et al. 1977; however, cf. Goldstein et al. 1976). The effects of opiates and SPA sum (Samanin and Valzelli 1971). Tolerance develops not only to opiates but also to SPA, and cross-tolerance can be demonstrated between these (Mayer and Hayes 1975; however, cf. Lewis and Gebhart 1977a,b).

Stimulation-produced analgesia in humans from stimulation of the PVG or PAG is associated with an increase in the concentration of met-enkephalin and β-endorphin in the CSF (Akil et al. 1978a,b, 1979; Hosobuchi et al. 1979). Cerebrospinal fluid endorphin levels are low in patients with chronic pain due to organic causes (Almay et al. 1978), and pain sensitivity in various clinical pain syndromes varies with CSF endorphin levels (Sjölund et al. 1977; Von Knorring et al. 1978) and with naloxone administration (Buchsbaum et al. 1977; Levine et al. 1978; however, cf. El Sobky et al. 1976; Grevert and Goldstein 1978; Lindblom and Tegner 1979). It is interesting that one form of congenital insensitivity to pain may involve a unusually active endogenous opiate system (Dehen et al. 1977; Thrush 1973). On the other hand, SPA from stimulation in the internal capsule does not result in changes in CSF β-endorphin levels (Hosobuchi et al. 1979).

3.3 Role of Monoamines in SPA

The neural systems for SPA and for opiate analgesia are complex, and no doubt involve a variety of chemical mediators. Apart from the endogenous opioid peptides, the transmitter candidates for which there is the most evidence are the monoamines: serotonin and the catecholamines. However, the exact role of monoamines in opiate analgesia is not yet at all clear, since there are many conflicting reports concerning the effects of various experimental manipulations. The evidence with respect to monoamine involvement in SPA will be considered first, since the few reports on this topic are reasonably consistent.

Stimulation-produced analgesia in rats due to stimulation in the PAG is reduced by treatment with tetrabenazine, which produces a nonspecific depletion of CNS monoamines (Akil and Liebeskind 1975). Analgesia is restored by administration of either 5-hydroxytryptophan (5-HTP) or L-dopa, the precursors respectively of serotonin and of the catecholamines (Akil and Liebeskind 1975). Specific depletion of serotonin by *p*-chlorophenylalanine (PCPA) also reduces SPA (Akil and Liebeskind 1975; Akil and Mayer 1972), as does depletion of catecholamines by α-methyl-*p*-tyrosine (AMPT) (Akil and Liebeskind 1975). Again, analgesia is restored by 5-HTP or by L-dopa, according to which monoamine is depleted (Akil and Liebeskind 1975). Furthermore, antagonism of serotonin transmission by lysergic acid diethylamide (LSD) also reduces SPA (Hayes et al. 1977a).

Disulfiram, a dopamine-β-hydroxylase (DBH) inhibitor, depletes norepinephrine without affecting dopamine levels. Treatment with disulfiram increases SPA, suggesting that norepinephrine tends to antagonize SPA. Haloperidol and pimozide, which block catecholamine receptors (especially dopamine receptors, in the case of pimozide), reduce SPA, and apomorphine, a dopamine agonist, increases SPA. These findings suggest that dopamine is involved in SPA and that norepinephrine interferes with SPA (Akil and Liebeskind 1975). A cautionary note comes from the work of Akil and Mayer (1972). In their experiments, PCPA reduced SPA only when the stimulus site was in the ventral part of the PAG. Stimulation-produced analgesia from points outside the PAG or from dorsal PAG was unaffected or was enhanced by PCPA.

Hosobuchi (1978; Hosobuchi et al. 1980) has found that patients being treated for intractable pain by implanted electrodes in the PVG or PAG may develop tolerance after months of such stimulation. The tolerance can be reversed by administration of tryptophan. The effect may result from replenishment of stores of serotonin in the CNS. A similar observation had previously been made with respect to animals in which a reduction of SPA due to prolonged stimulation of the NRM was reversed by 5-HTP administration (Oliveras et al. 1978).

Stimulation-produced analgesia from stimulation in the lateral hypothalamus is reported to be unaffected by PCPA or by AMPT (Yunger et al. 1973). However, this form of SPA is likely to differ from that evoked from the PAG (Akil and Liebeskind 1975). Recently, Satoh et al. (1980a) found that SPA from the nucleus paragigantocellularis is antagonized by an α-adrenergic blocker (as well as by naloxone), whereas SPA from the NRM is antagonized by a serotonin receptor blocker (but not by naloxone). They suggest that there are two pathways from the brain stem mediating analgesia, one utilizing norepinephrine and the other serotonin.

More complicated arrangements are also likely. For example, microinjection of the α-adrenergic antagonist phentolamine into the NRM causes analgesia that is reversed by intrathecal administration of the serotonin antagonist methysergide (Hammond et al. 1980). Their findings are interpreted to indicate that there is a tonic inhibition of serotonergic raphe-spinal neurons by an α-adrenergic mechanism. Intrathecal phentolamine also partially reverses the analgesia, suggesting that noradrenergic projections are also involved (Sagen and Proudfit 1981).

3.4 Role of Monoamines in Opiate Analgesia

There is considerable evidence that CNS monoaminergic synapses play a role in the production of analgesia by opiates like morphine (see, e.g., Way 1972). The evidence from pharmacological studies includes observations of the effects of depletion of monoamines, blockade or stimulation of monoamine receptors, changes in monoamine turnover, and release or blockade of reuptake of monoamines on opiate analgesia. Ancillary evidence comes from studies of the actions of drugs acting at monoamine synapses or activation of monoamine-containing neurons on pain responses.

Like SPA, opiate analgesia can be altered by depletion of monoamines. For example, several investigators report that reserpine antagonizes opiate analgesia [Fennessy and Lee 1970; Grossman et al. 1973; Major and Pleuvry 1971; Medakovic and Banic 1964; Radouco-Thomas et al. 1957; Ross and Ashford 1967 (tail clip test); Rudzik and Mennear 1965; Schaumann 1958; Takagi et al. 1964; Vedernikov and Afrikanov 1969]. Since reserpine nonselectively depletes serotonin and the catecholamines, the effects of replenishment of different amines by precursors were tried. 5-Hydroxytryptophan reversed the effect of reserpine, whereas L-dopa did not (Major and Pleuvry 1971). However, others find that morphine analgesia is either unaffected by reserpine (Buxbaum et al. 1973) or is enhanced [Garcia and Rocha e Silva 1961; Ross and Ashford 1967 (hot plate test)]. Tetrabenazine also antagonizes morphine analgesia (Takagi et al. 1964; Takagi and Nakama, 1968). Specific depletion of serotonin also results in different effects on morphine analgesia, depending upon the particular laboratory or the analgesia tests used. Morphine analgesia is reduced by PCPA or by the neurotoxin 5,6-dihydroxytryptamine, according to some [Fennessy and Lee 1970 (phenylquinone writhing test); Genovese et al. 1973; Görlitz and Frey 1972; Major and Pleuvry 1971; Takemori et al. 1975; Tenen 1968; Tulunay et al. 1976; Vogt 1974]. According to others, there is no change after PCPA [Bläsig et al. 1973; Fennessy and Lee 1970 (hot plate test); Buxbaum et al. 1973; Cheney and Goldstein 1971; Reinhold et al.

1973] or an increase (Saarnivaara 1969a). Some of these discrepant findings may be due to species differences, to the analgesia test employed or to the dose of morphine used (Vogt 1974; York and Maynert 1978). When PCPA reduces morphine analgesia, the analgesia may be restored by 5-HTP (Görlitz and Frey 1972; Major and Pleuvry 1971). However, Ho et al. (1975) find that tryptophan antagonizes morphine analgesia.

Specific depletion of catecholamines by AMPT, by inhibition of DBH, or by administration of a neurotoxin may either reduce morphine analgesia [Ayhan 1972; Görlitz and Frey 1972; Major and Pleuvry 1971 (inhibition of DBH); Medakovič and Banič 1964 (mice); Nakamura et al. 1973 (dopamine in caudate); Reinhold et al. 1973; Saarhivaara 1969b], have no effect [Fennessy and Lee 1970; Medakovič and Banič 1964 (rats); York and Maynert 1978 (tail flick)], or enhance [Buxbaum et al. 1973; Cicero et al. 1974; Major and Pleuvry 1971 (AMPT); Nakamura et al. 1973 (norepinephrine in hypothalamus); York and Maynert 1978 (squeal threshold)] morphine analgesia. When morphine analgesia is reduced by catecholamine depletion, it may be restored by administration of L-dopa (Ayhan 1972; Saaraivaara 1969b). Furthermore, when tetrabenazine is used to produce a nonspecific depletion of catecholamines and serotonin, the reduction in morphine-induced analgesia can be counteracted by administration of dopa (Takagi et al. 1968).

Blocking agents have also been tried on morphine analgesia. Serotonin blockers, such as methysergide or cyproheptadine, may antagonize morphine analgesia (Görlitz and Frey 1972), have no effect (Fennessy and Lee 1970), or enhance morphine analgesia (Saarnivaara 1969a). Similarly, α-adrenergic blockers, such as phenoxybenzamine and phentolamine, have no effect on (Fennessy and Lee 1970; Görlitz and Frey 1972; Takemori et al. 1975; Tulunay et al. 1976) or enhance (Cicero et al. 1974) morphine analgesia. Chlorpromazine, haloperidol, and pimozide can antagonize morphine analgesia (Görlitz and Frey 1972; Tulunay et al. 1976). β-Adrenergic blockers like propranolol have no effect on morphine analgesia (Fennessy and Lee 1970; Cicero et al. 1974; Görlitz and Frey 1972; Takemori et al. 1975; Tulunay et al. 1976), although they can potentiate the antagonistic action of naloxone (Takemori et al. 1975).

Other pharmacological manipulations suggest a role for the monoamines in analgesia. For example, both amphetamine and *p*-chloroamphetamine are reported to cause analgesia (Görlitz and Frey 1972; Vedernikov 1969). These agents are known to cause release, respectively, of catecholamines and serotonin. Furthermore, the analgesia produced by administration of amphetamine and *p*-chloroamphetamine is abolished by pretreatment with PCPA (Görlitz and Frey 1972). Depletion of serotonin by PCPA does not affect the threshold of the responses of rats to foot shock but does increase the responses to superthreshold stimuli (Fibiger et al. 1972). Diets deficient in tryptophan result in hyperalgesia that can be corrected by tryptophan replacement or by pharmacological treatment with fluoxetine (Messing et al. 1976).

Tryptophan or 5-HTP administration is reported to reverse tolerance to opiates in humans (Hosobuchi et al. 1980) and in animals (Oliveras et al. 1978). Fluoxotine, a serotonin uptake blocker, causes analgesia that sums with that produced by morphine; naloxone does not reverse fluoxetine analgesia (Messing et al. 1975).

Dopamine receptor agonists, like apomorphine and L-dopa, can reduce morphine analgesia (Major and Pleuvry 1971; Takemori et al. 1975; Tulunay et al. 1976; Vedernikov 1969), but L-dopa has also been reported to elevate pain threshold and tolerance to radiant heat stimuli (Battista and Wolff 1973).

Morphine produces an increased turnover of CNS serotonin (Görlitz and Frey 1972; Goodlet and Sugrue 1974; Haubrich and Blake 1973; Yarbrough et al. 1972, 1973), norepinephrine [Görlitz and Frey 1972; Sugrue 1974 (brain stem, not hypothalamus)], and dopamine (Sugrue 1974; however, cf. Görlitz and Frey 1972). While an increased turnover of monoamines is consistent with their involvement in morphine analgesia, this by no means proof, since the monoamines could as well play a role in the other actions of morphine (Yarbrough et al. 1973). For example, it has been proposed that alterations of monoaminergic transmission result in changes in opiate analgesia tests not because of an action on the analgesia mechanism per se, but because of an action of the motor system (cf. Buxbaum et al. 1973; Cheney and Goldstein 1971; Grossman et al. 1973; Yarbrough et al. 1973). However, it is difficult to distinguish between an action on "analgesia" and one on the motor system so long as "analgesia" is judged on the basis of motor responses.

Monoamines, including serotonin and norepinephrine, are released into the spinal CSF during analgesia produced by morphine injections into the brain or by stimulation of nervous tissue (Yaksh 1979; Yaksh and Tyce 1979), and analgesia is produced by intrathecal administration of serotonin or norepinephrine (Reddy and Yaksh 1980; Reddy et al. 1980; Wang 1977; Yaksh and Wilson 1979). The analgesia produced by microinjection of morphine into the PAG can be partially antagonized by intrathecal administration of either a serotonin or an α-adrenergic antagonist, methysergide or phentolamine (Fig. 21; Yaksh 1979; Yaksh et al. 1976a). The action of norepinephrine appears to be mediated by α- and not by β-receptors (Reddy and Yaksh 1980; Reddy et al. 1980). Furthermore, analgesia is produced by stimulation in brain nuclei that contain serotonin or norepinephrine (Bowker et al. 1981; Dahlström and Fuxe 1965; DiCarlo et al. 1974; Hubbard and DiCarlo 1974; Felten et al. 1974) and that project to the spinal cord (Besson et al. 1981; Dickenson et al. 1979; Margalit and Segal 1979; Oliveras et al. 1975, 1978, 1979; Sandberg and Segal 1978; Segal and Sandberg 1977). However, at least part of the action of monoamines on nociception is likely to be at the level of the brain, rather than in the spinal cord (Harvey and Lints 1971; Price and Fibiger 1975; Yaksh 1979; Yunger and Harvey 1976).

Recently, it has been found that intrathecal administration of a serotonin blocker, methysergide, or an α-adrenergic blocker, phentolamine, produces hyperalgesia (Proudfit and Hammond 1981). However, intrathecal methysergide did not alter the analgesia produced by systemic administration of morphine. Phentolamine on the other hand, antagonized the effect of a low dose of morphine, although not the effect of a high dose. It was concluded that morphine analgesia does not depend upon the release of serotonin by raphe-spinal neurons, although it may be mediated in part by descending noradrenergic neurons. However, there is a tonic inhibition of spinal cord nociceptive transmission by serotonergic and noradrenergic paths.

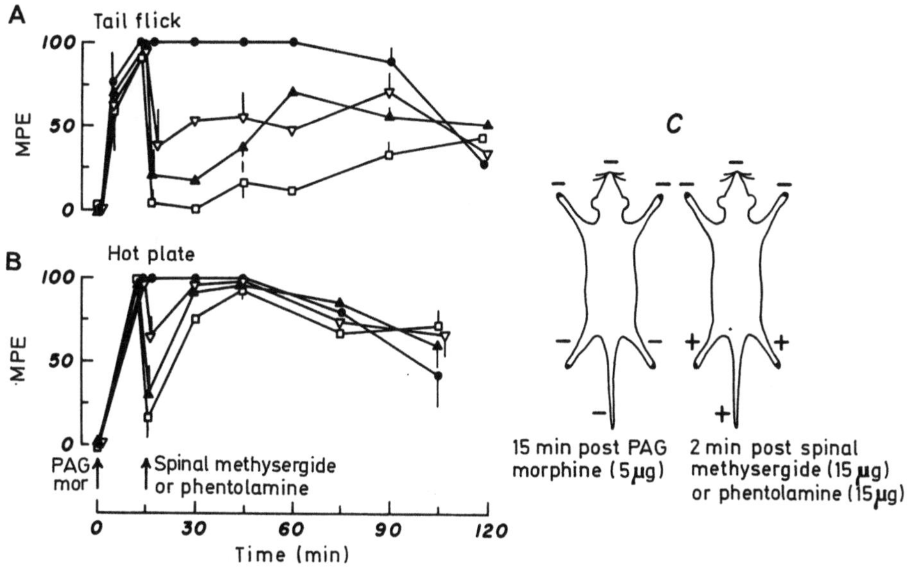

Fig. 21A – C. Interaction between morphine analgesia and intrathecal serotonin or α-adrenergic blockers. Analgesia was judged by the tail flick (**A**) and hot plate (**B**) tests. Morphine (5 μg) was injected into the periaqueductal gray (*PAG*) at the time of the *first arrow*. At the *second arrow*, intrathecal saline (●), methysergide (▽), phentolamine (▲), or methysergide plus phentolamine (□) was given. In **C**, the *figurine at the left* indicates that morphine blocked the responses to pinching the places shown on the body and face with forceps. The *figurine at the right* indicates that blockade of serotonin or α-adrenergic receptors prevented the morphine analgesia on the hind limbs and tail when administered intrathecally, but not the analgesia on the forelimbs and face. (Yaksh 1979)

3.5 Other Candidate Transmitters and Modulators

Cholinergic mechanisms may play a part in opiate analgesia. It has been reported that muscarinic blockers, such as atropine, reduce morphine analgesia, whereas cholinesterase inhibitors, such as physostigmine, enhance morphine analgesia (Bhargara and way 1972; Flodmark and Wramner 1945; Ireson 1970; Takemori et al. 1975; Tulunay et al. 1976). Cholinomimetic drugs can themselves produce analgesia (Flodmark and Wramner 1945; Ireson 1970).

3.6 Pharmacology of Analgesia Due to Acupuncture and Transcutaneous Nerve Stimulation

The role of the endogenous opioid substances in acupuncture analgesia has been discussed in Section 2.4. It is currently thought that the serotonergic pathways of the CNS are also involved in acupuncture analgesia (Cheng and Pomeranz 1981; Chiang et al. 1979; Han et al. 1979, 1980; Kin et al. 1979; McLennan et al. 1977), and acupuncture causes release of radioactively labelled serotonin into the ventricular system (Yi et al. 1977). A similar statement can be made about the acupuncture-like form of transcutaneous nerve stimulation (Woolf et al. 1980; see

Fig. 15). On the other hand, adrenergic mechanisms seem to be important for the analgesia in rats produced by vaginal stimulation (Crowley et al. 1977b). Furthermore, inhibitory mechanisms involving glycine and γ-aminobutyric acid are also likely to be involved, since acupuncture analgesia is partially antagonized by strychnine and by bicuculline (McLennan et al. 1977). Similarly, cholinergic mechanisms have been implicated, because acupuncture analgesia is antagonized by cholinergic blocking agents (Han et al. 1980). Adrenergic mechanisms play a more complex role, which is incompletely understood (Han et al. 1980).

3.7 Conclusions

1. The injection of morphine into the brain results in analgesia. The most sensitive sites for morphine administered into the cerebrospinal fluid are near the rostral fourth ventricle.
2. Microinjections of morphine (μg quantities) into the periaqueductal and periventricular gray produce analgesia. The same locations when stimulated electrically sometimes, but not always, result in stimulation-produced analgesia. The analgesia due to microinjected morphine is reversed by the opiate antagonist, naloxone.
3. Microinjections of morphine into the nucleus paragigantocellularis lateralis of the rat medulla in ng quantities also produce a naloxone-reversible analgesia. Microinjections of morphine into the nucleus raphe magnus can cause analgesia.
4. Systemically administered morphine is thought by some to act chiefly on opiate receptors in the brain, although, especially with higher doses, an action on the spinal cord is also likely. A selective action on the spinal cord is possible by intrathecal administration of morphine.
5. It has been proposed that the neural pathways mediating stimulation-produced and opiate analgesia are similar and involve the midbrain periaqueductal gray, a synaptic link in the nucleus raphe magnus and perhaps the adjacent reticular formation, and axons in the dorsal parts of the lateral funiculi that terminate in the upper layers of the spinal cord gray matter.
6. Stimulation-produced analgesia in humans is associated with the release of opioids when the stimulus sites are in the periventricular or periaqueductal gray. Cerebrospinal fluid endorphin levels appear to reflect clinical pain states.
7. Stimulation-produced analgesia involves monoamines as well as opioid substances. A role of serotonin and of catecholamines is suggested by pharmacological manipulations that involve depletion, antagonistic or agonistic actions, or administration of monoamine precursors after depletion. Serotonin and dopamine may be concerned with the production of analgesia and norepinephrine with interference with analgesia.
8. Opiate analgesia is also thought to involve monoamine transmitters. Depletion of serotonin or of catecholamines reduces morphine analgesia, according to many, but not all, investigators. Monoamine antagonists and agonists have

predictable effects in reducing or enhancing morphine analgesia, and agonists or releasing agents may produce analgesia in their own right. Morphine causes an enhanced turnover of monoamines, and when microinjected into the periaqueductal gray causes the release of monoamines in the spinal cord.

9. Cholinergic mechanisms may also be involved in opiate analgesia.
10. Monoaminergic and cholinergic mechanisms have been implicated in acupuncture analgesia. Amino acid inhibitory transmitters may also be involved.

4 Descending Control of the Flexion Reflex

4.1 Organization of the Flexion Reflex

The flexion reflex was one of the simple reflexes that Sherrington used to illustrate many of the principles of reflex action that he described in *The Integrative Action of the Nervous System* (Sherrington 1906). Particularly important features of the flexion reflex are that: (a) the receptive field includes both the skin and muscle; (b) the reflex may "irradiate" into other limbs, e.g., leading to the crossed extension reflex in the contralateral limb and extension and flexion in the other two limbs (Fig. 22A); and (c) the flexion reflex may supplant other, less dominant reflexes (Sherrington 1906). A series of papers on the flexion reflex was published by Sherrington's laboratory (Cooper et al. 1927; Creed and Sherrington 1926; Eccles 1931; Eccles and Granit 1929; Eccles and Sherrington 1930, 1931a,b,c,d,e; Liddell and Sherrington 1923; Sherrington 1910) and by several American colleagues (Forbes and Cattell 1924; Forbes and Gregg 1915; Pi-Suner and Fulton 1929). Much of the information from these studies is summarized in *Reflex Activity of the Spinal Cord* (Creed et al. 1932).

The flexion reflex is readily elicited in the lower extremities of an animal whose spinal cord has been transected. A needle prick may elicit a flexor withdrawal response in the hind limb of a chronically spinalized dog suspended off the ground. The reflex may continue for several seconds. Destruction of the spinal cord or the dorsal or ventral roots abolishes the reflex. The adaptive function of the reflex is "to withdraw the limb from contact with injurious agents" (Creed et al. 1932). After making allowances for nerve conduction time, the minimum CNS delay in the flexion reflex pathway is estimated to be about 3 – 5 ms (Creed et al. 1932). However, the delay for weak responses is much longer (Fig. 22B; Creed et al. 1932). The flexion reflex varies in amplitude and duration according to the strength of stimulation (Creed et al. 1932). The cause of the prolonged contraction of flexor muscles that results from a strong stimulus could be either a prolonged input due to repetitive discharges in nociceptive afferents or a long-lasting central action. These alternatives were tested by eliciting the flexion reflex with electric shocks applied to a peripheral nerve (Fig. 22B). An electrical stimulus results in an afferent discharge only at the time of stimulation, yet a prolonged contraction still follows a strong stimulus. Therefore, the prolonged contraction of the flexion reflex can be attributed to after-discharges of central neurons (additional afferent input due to the reflex contraction can be ruled out as the sole explanation by the fact that a prolonged contraction can also be elicited in a deafferented preparation, according to Eccles and Sherrington 1930). Repetitive stimulation can produce a tetanic contraction, although this is not always maintained during continued stimulation (Creed et al. 1932).

The flexion reflex in a hind limb involves contractions of flexor muscles at the ankle, knee, and hip (Creed et al. 1932). The forces of contraction of the different flexor muscles will vary, however, according to the source of the afferent input. This is the basis of the "local sign" of the flexion reflex, the variation in

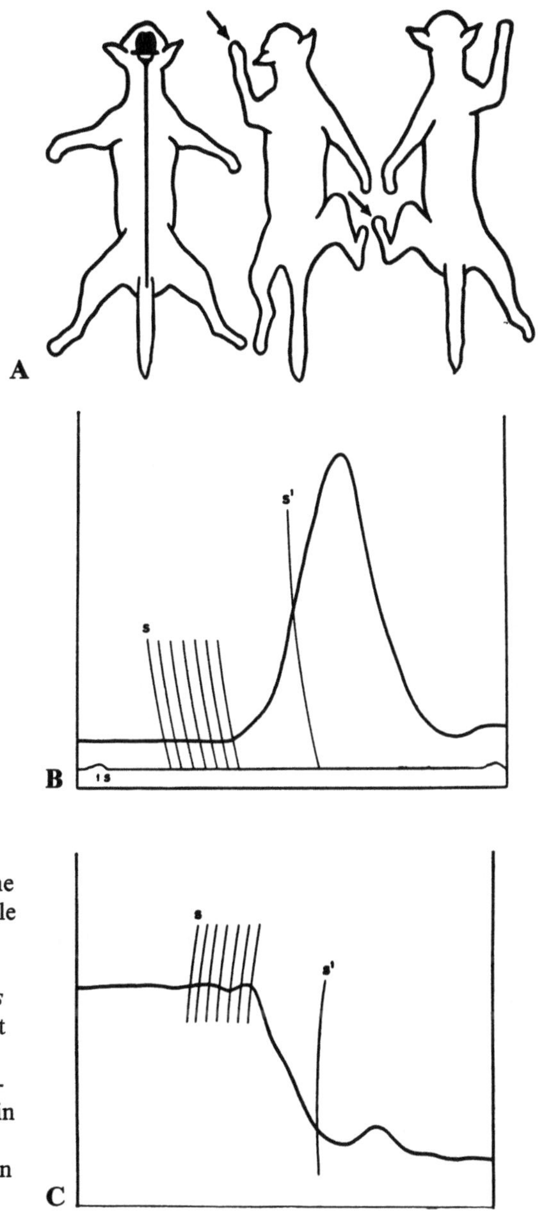

Fig. 22A – C. Flexion reflex in the cat. **A** *Left*, posture of decerebrate cat. Changes in position due to stimulation of left forepaw or left hind paw shown in *center* and at *right*. **B** Myographic recording of the contraction of the knee flexor muscle in response to electrical stimulation of branch of the saphenous nerve, using a train of shocks. *Time marks* show 1 s. *S* and *S1* indicate the start and end of the stimulus train. **C** Myographic recording of the relaxation of the knee extensor muscle in response to stimulation of a branch of the saphenous nerve, using a train of shocks. (Sherrington 1906)

the details of the reflex according to the point of application of the noxious stimulus (Sherrington 1906).

Accompanying the contraction of flexor muscles during the flexion reflex is an inhibition of the antagonistic extensor muscles (Fig. 22C; Sherrington 1906). This inhibition is one of the instances noted by Sherrington of reciprocal inhibition as a principle of reflex organization. Another example is the crossed extension reflex which may accompany the flexion reflex. As mentioned earlier, the flexion reflex can irradiate into the contralateral limb, resulting in a contraction

Fig. 23. Interruption of reflex stepping of the hind limb of a decapitate cat by a nociceptive flexion reflex (myographic recording, with downward deflection during flexion). (Sherrington 1910)

of the extensor muscles and a relaxation of the flexor muscles (Creed et al. 1932; Sherrington 1906).

These early investigations of the flexion reflex in the Sherrington era have so far been characterized for the case of the flexor withdrawal reflex that results from a noxious stimulus. However, in addition to this reflex, there should be mentioned another flexion reflex which contributes to locomotion (Brown 1911, 1912, 1914; Sherrington 1910; Engberg and Lundberg 1969). Although there are many similarities in these two forms of the flexion reflex, their separate adaptive functions lead to important differences. The most obvious difference is that the nociceptive flexion reflex interrupts the alternation of flexion and extension that characterizes locomotion (Fig. 23; Sherrington 1910). That is, the flexor withdrawal reflex supersedes the flexion and extension reflexes of locomotion. Another difference is the higher intensity of stimulation required to elicit the flexor withdrawal reflex. Reflex stepping is best elicited by innocuous, proprioceptive stimuli, whereas the flexor withdrawal reflex is evoked by noxious stimuli. Furthermore, the nociceptive flexion reflex is simultaneous at all joints, whereas flexion during stepping is not, since during locomotion different joints show a cessation of flexion at different times (Engberg and Lundberg 1969). The nociceptive flexion is succeeded by extension only when the noxious stimulation is halted (Sherrington 1910).

The modern analysis of the flexion reflex began with the work of Lloyd (1943a), who demonstrated that the discharges of motoneurons participating in the flexion and other reflexes could be recorded from the ventral roots of acutely prepared animals (cf. Forbes and Gregg 1915). When a dorsal root is stimulated, the reflex discharge recorded from the ventral root of the same segment is complex, including an early synchronous component and a later, asynchronous, discharge. The initial discharge is produced nearly in isolation when a muscle nerve is stimulated, and the later discharge is produced alone when a cutaneous nerve is stimulated (Fig. 24; Lloyd 1943a,b). The early, synchronous, discharge produced by stimulating a muscle nerve is due to monosynaptic excitation of motoneurons by group Ia afferent fibers from muscle spindle afferents (Alvord

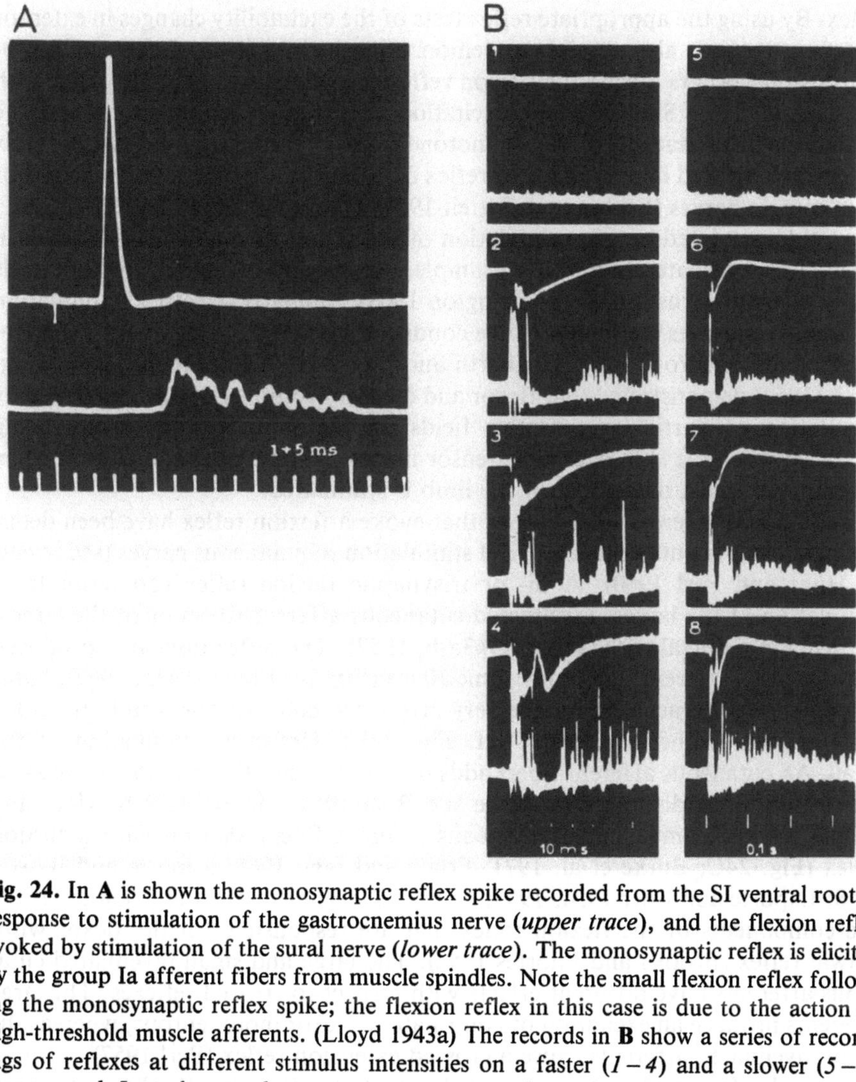

Fig. 24. In **A** is shown the monosynaptic reflex spike recorded from the SI ventral root in response to stimulation of the gastrocnemius nerve (*upper trace*), and the flexion reflex evoked by stimulation of the sural nerve (*lower trace*). The monosynaptic reflex is elicited by the group Ia afferent fibers from muscle spindles. Note the small flexion reflex following the monosynaptic reflex spike; the flexion reflex in this case is due to the action of high-threshold muscle afferents. (Lloyd 1943a) The records in **B** show a series of recordings of reflexes at different stimulus intensities on a faster (*1 – 4*) and a slower (*5 – 8*) sweep speed. In each case, the *upper traces* are recordings of the afferent volley from the dorsal root, and the *lower traces* are from the nerve to semitendinosus. There was no stimulus in *1* and *5*. For *2* and *6*, the stimulus was maximal for A fibers in the dorsal root. For *3* and *7* and for *4* and *8*, two different intensities were used for activation of part or all of the C fibers. (Laporte and Boër 1955)

and Fuortes 1954; Eccles et al. 1957; Lloyd 1943b,c; Lloyd and Chang 1948; Renshaw 1940; Rexed and Therman 1948; Lloyd 1946b), whereas the later, asynchronous, discharge resulting from stimulation of a cutaneous nerve (or of high-threshold muscle or joint afferents) is due to a disynaptic or polysynaptic excitation of the motoneurons to flexor muscles (Alvord and Fuortes 1954; Lloyd 1943b, 1946a,b, 1957). The polysynaptic discharge thus represents the motoneuronal output command that causes the reflex contractions of the flexion reflex. By using the appropriate reflex tests of the excitability changes in extensor motoneurons, it is also possible to demonstrate the inhibition of extensor motoneurons that occurs during the flexion reflex (Lloyd 1946a,b; see also Hagbarth and Naess 1950). Similarly, the excitation of extensor motoneurons and the inhibition (or excitation) of flexor motoneurons in the crossed extension reflex can be documented in recordings of reflex discharges either from ventral roots or from muscle nerves (Forbes and Cattell 1924; Holmqvist 1961; Perl 1957).

It should be pointed out the stimulation of the skin does not invariably result in flexion of the ipsilateral limb. For example, pressure on the skin of the foot leads to the extensor thrust reflex (Sherrington 1903). Similarly, electrical stimulation of cutaneous nerves can, under some conditions, lead to activation of ipsilateral extensor motoneurons (e.g., Hagbarth and Naess 1950; Rosenberg 1970). Hagbarth (1952) demonstrated that flexor and extensor responses could be elicited by stimulation of particular receptive fields on the skin, with extension being produced when the skin over an extensor muscle is stimulated and flexion when the skin over most of the rest of the limb is stimulated.

The cutaneous afferent fiber classes that evoke a flexion reflex have been determined in experiments using electrical stimulation of cutaneous nerves (see review by Hunt and Perl 1960). A di- or trisynaptic flexion reflex can result from stimulation of the largest myelinated cutaneous afferent fibers or of the largest afferents in a dorsal root (Lloyd 1943a,b, 1957). The reflex discharge produced by these large afferents lasts only some 10 ms (Fig. 24; Lloyd 1943a, 1957), hardly long enough to account for the very prolonged contractions seen in the nociceptive flexor withdrawal reflex (cf. Fig. 22B). However, stimulation of the small, Aδ cutaneous afferent fibers adds both to the duration and the intensity of the flexion reflex discharge (Laporte and Boër 1954; Lloyd 1943a,b; Mark and Steiner 1958). Unmyelinated cutaneous afferent fibers also produce a flexion reflex (Fig. 24C; Burke et al. 1971; Franz and Iggo 1968; Laporte and Bessou 1958; Laporte and Boer 1954, 1955; Price 1972; Voorhoeve et al. 1958). It is the late components of the flexion reflex that are responsible for the flexor withdrawal reflex (Brooks and Fuortes 1952). The large and small myelinated cutaneous afferents have different effects with respect to crossed reflexes. The large afferents produce an excitation of contralateral flexor motoneurons, whereas the small myelinated afferents cause a crossed extension reflex (Perl 1957).

The muscle afferent fiber groups contributing to the flexion reflex have also been studied using electrically evoked volleys. Group II muscle afferent fibers, presumably of muscle spindle origin (Hunt 1954), have been found to have flexion reflex actions, producing an ipsilateral flexion (Eccles R. M. and Lundberg 1959a; Kuno and Perl 1960; Laporte and Lloyd 1952; Lloyd 1943b, 1946b), as well as crossed extension (Perl 1958). However, there is more recent evidence supporting the notion that group II muscle spindle afferent fibers participate in the stretch reflex (Kirkwood and Sears 1974, 1975; Lundberg et al. 1977; Matthews 1969, 1973; Stauffer et al. 1976; cf. Cangiano and Lutzemberger 1972; Grillner 1970). The group II muscle afferents that play a role in the flexion reflex may be pressure-pain endings (Paintal 1960), rather than muscle spindle afferents (Matthews 1969). Group III muscle afferent fibers also contribute to the flexion reflex (Hagbarth and Naess 1950; Lloyd 1943b; Paintal 1961). Crossed effects include short-lasting facilitation of both flexor and extensor motoneurons and long-lasting facilitation of extensor motoneurons (Perl 1958). Since group III muscle afferent fibers supply nociceptors (Mense 1977; Paintal 1960), the flexion reflex elicited by stimulation of group III muscle afferents may well be part of the nociceptive flexor withdrawal reflex.

High-threshold joint afferents also have flexion reflex actions (Eccles R. M. and Lundberg 1959b; Ekholm et al. 1960; Gardner 1950; Grigg et al. 1978). Large afferents in joint nerves may facilitate extensor and inhibit flexor motoneurons, although the actions are often weak and variable (Grigg et al. 1978; Skoglund 1956). However, joint afferents are quite effective in modulating the activity of the pathways for the group Ib and flexion reflexes (Baxendale and Ferrell 1981; Lundberg et al. 1978).

The various types of primary afferent fibers that evoke the flexion reflex have been lumped together by Lundberg and his associates in a broad category they call the "flexion reflex afferents" (FRA) (Eccles R. M. and Lundberg 1959a; Holmqvist 1961; Holmqvist and Lundberg 1961). These include cutaneous myelinated afferents of all sizes and small myelinated muscle and joint afferents. For some reason, visceral afferents and unmyelinated fibers were left out of the FRA classification. Although convenient, the term FRA has not been universally accepted (see Boivie and Perl 1975; Matthews 1972; McIntyre 1974; Stauffer et al. 1976; Wilson and Kato 1965).

After the intracellular recording technique was introduced, the mechanism of excitation and inhibition of motoneurons during the flexion reflex was investigated further in the laboratories of Eccles, Lundberg, and others. The excitation of flexor motoneurons by polysynaptic excitatory postsynaptic potentials mediated by spinal cord interneurons explains the latency and form of the flexion reflex (Fig. 25; Eccles R. M. and Lundberg 1959a,b). Simultaneously, extensor motoneurons generate inhibitory postsynaptic potentials which largely explain

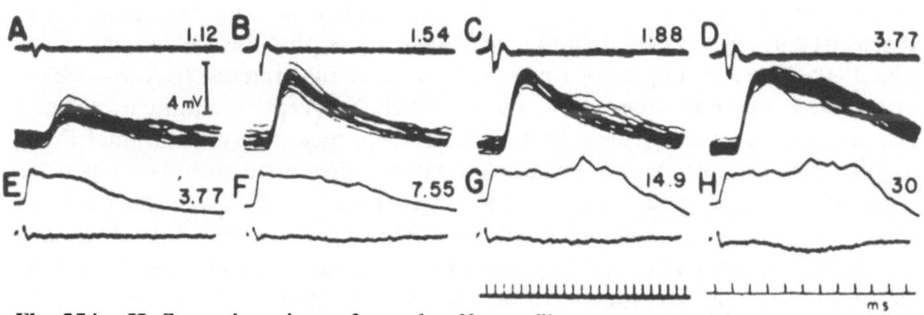

Fig. 25A – H. Synaptic actions of muscle afferent fibers as recorded from a flexor motoneuron. The *upper traces* in **A – D** and the *lower traces* in E – H show the afferent volleys and cord dorsum potentials recorded from the surface of the spinal cord. The *lower traces* in **A – D** and *upper traces* in **E – H** are intracellular records from a motoneuron supplying the biceps-semitendinosus muscles. The biceps-semitendinosus nerve was stimulated at progressively increasing strengths in **A – H**; stimulus strength is expressed in multiples of threshold for the largest afferent fibers above each record. **A – D** were at a faster and E – H at a slower sweep speed. The excitatory postsynaptic potential in **A – C** was a monosynaptic one due to activation of group Ia afferent fibers, whereas the longer-latency excitatory potentials in the other records were due to group II and at higher strengths group III afferent fibers. (R. M. Eccles and Lundberg 1959a)

the reciprocal inhibition of these cells during the flexion reflex (Eccles R. M. and Lundberg 1959b; Rosenberg 1970). The effective afferents include group II and III muscle afferents and skin and joint afferents (Eccles R. M. and Lundberg 1959a,b; Rosenberg 1970; Wilson and Kato 1965). However, primary afferent depolarization also accompanies the flexion reflex and undoubtedly presynaptic inhibition contributes to the inhibitory events (Burke et al. 1971; Eccles et al. 1962, 1963).

Some classes of extensor motoneurons show mixtures of excitatory and inhibitory postsynaptic potentials during the flexion reflex (Burke et al. 1970; Rosenberg 1970). Furthermore, the sign of the dominant synaptic potential evoked by high-threshold muscle afferents may vary with the state of the animal preparation (Eccles R. M. and Lundberg 1959a). The particular mixture depends upon the motoneuronal type, with inhibitory postsynaptic potentials being more prominent in slow twitch motor units and with a trend toward more excitation in fast twitch units (Burke et al. 1970). It should be determined whether there is also a relationship to the cutaneous receptive field location (cf. Hagbarth 1952).

The interneurons that are activated by primary afferent fibers evoking the flexion reflex have not been satisfactorily identified. A detailed identification of the first-, second-, and higher-order interneurons in the pathways from the terminals of particular classes of afferent fibers to the excitatory and inhibitory synaptic terminals on flexor and extensor motoneurons has not as yet been accomplished, although considerable progress has been made for less complex interneuronal pathways, such as the group Ia reciprocal inhibitory path (Jankowska and Roberts 1972a,b; Jankowska and Lindström 1972), and the recurrent inhibitory path (Jankowska and Lindström 1971). Maps are available of the locations of interneurons that are excited by electrical stimulation of cutaneous or muscle afferents (e.g., Armett et al. 1961; Cervero et al. 1976; Christensen and Perl 1970; Eccles et al. 1960; Fukushima and Kato 1975; Gregor and Zimmermann 1972; Handwerker et al. 1975; Hongo et al. 1966; Kolmodin and Skoglund 1960; Menétrey et al. 1977; Wagman and Price 1969). Given that the flexor withdrawal reflex is triggered by noxious stimuli, it is of particular interest to know where nociceptive spinal cord interneurons are located. Nociceptive interneurons have been found in several locations in the dorsal horn, including the marginal zone, the substantia gelatinosa, and the neck and base of the dorsal horn (laminae I, II, V, and VI of Rexed 1952; see Section 5; Bennett et al. 1979, 1980; Besson et al. 1972; Cervero and Iggo 1980; Cervero et al. 1976, 1977b, 1979; Christensen and Perl 1970; Guilbaud et al. 1977b; Handwerker et al. 1975; Hillman and Wall 1969; Kolmodin and Skoglund 1960; Kumazawa and Perl 1978; Kumazawa et al. 1975; Light et al. 1979; Menétrey et al. 1977; Price et al. 1979; Price and Mayer 1974; Wall 1967; Wall et al. 1979). Although many of the cells in these regions are propriospinal neurons, it is not known if any project directly to motoneurons. A crucial problem is whether such neurons are part of the flexion reflex pathway or part of a sensory circuit transmitting information to the brain. It is of course possible for dorsal horn interneurons to participate both in reflex and in sensory mechanisms.

Since the FRA include all sizes of myelinated cutaneous afferents, there should be spinal cord interneurons that respond both to innocuous and noxious stimulation of the skin. In fact, such cells form the largest class of nociceptive interneurons in the dorsal horn. These nociceptive cells are often now called "wide dynamic range" cells, with the identifying characteristic being that they respond to a wide range of stimulus intensities and to both A and C fiber volleys (Mendell 1966; Wall 1967). Many wide dynamic range interneurons are excited not only by cutaneous afferents but also by muscle and visceral afferents (Pomeranz et al. 1968). Thus their input seems to include not only the fibers defined by Lundberg as FRA, but also several fiber groups not included in the FRA (visceral afferents; C fibers) but which are responsible for evoking flexion reflexes.

In addition two wide dynamic range cells, the dorsal horn contains cells that are not activated by innocuous stimuli but that are activated by noxious stimuli. These cells are often termed "nociceptive-specific" or "high-threshold" cells, and they are particularly abundant in the marginal zone and substantia gelatinosa (laminae I and II; Cervero et al. 1976; Christensen and Perl 1970; Kumazawa et al. 1975; Menétrey et al. 1977; Price et al. 1979; Price and Mayer 1974), although similar cells are also found in deeper layers of the dorsal horn.

One is tempted to exclude from the flexion reflex pathways interneurons that have been identified as the linkages in disynaptic pathways, such as the group Ia reciprocal inhibitory pathway or the group Ib pathway. However, these interneurons have been shown to receive polysynaptic actions from the same afferents that produce the flexion reflex (Eccles et al. 1960; Hongo et al. 1966). Perhaps these interneurons serve as higher-order cells in the flexion reflex pathways, in addition to being second-order interneurons in the group I pathways.

Little is known about the flexion reflex pathways in the rat spinal cord. This is unfortunate, since the tail flick test for analgesia is a flexor withdrawal reflex. More is known about the jaw-opening reflex, another analgesia test and a flexion reflex of the trigeminal system. Interestingly, there are two forms of jaw-opening reflex, one elicited by noxious stimulation and another by innocuous stimuli (Kidokoro et al. 1968; McGrath et al. 1981; Pfaffman 1939; Sessle and Schmitt 1972; Sherrington 1917; Thexton 1974). It seems likely that the jaw-opening reflex associated with innocuous inputs from oral mechanoreceptors is involved in mastication, whereas that due to noxious inputs is equivalent to a flexor withdrawal reflex (see Dubner et al. 1978). Thus the jaw-opening reflex and the flexion reflex are both dual in nature, having a low-threshold component that contributes to a rhythmic response (chewing, locomotion), and a high-threshold component serving a protective role.

4.2 Descending Pathways That Control the Flexion Reflex

The first clear demonstration that the flexion reflex is subject to inhibitory control by pathways descending from the brain was by Sherrington and Sowton (1915), who found that the flexion reflex evoked by electrical stimulation of a peripheral nerve in a decerebrate animal is enhanced following spinal cord transection. This finding was confirmed by Forbes et al. (1923) and by Liddell et al. (1932). Liddell et al. found in cats that tracts both in the dorsolateral and the ventral white matter contribute to the inhibition, whereas Chambers et al. (1970) showed that in monkeys the inhibition is mediated via the dorsal parts of the lateral funiculi. The idea that this change is due to the removal of a tonic inhibitory influence on the interneurons involved in the flexion reflex pathway can be attributed to Fulton (1926). Evidence supporting this concept came from later investigations showing that stimulation in the reticular formation can cause a reduction in the cord dorsum potentials and in polysynaptic reflexes (Austin 1952; Kleyntjens et al. 1955; Lindblom and Ottoson 1956), and that spinal cord transection results in an increased amplitude of the cord dorsum N wave (Lindblom and Ottosson 1953). Modern confirmation of the observation of Sherrington and Sowton (1915) was provided by Job (1953), R. M. Eccles and Lundberg (1959b), and Kuno and Perl (1960). In the context of pain mechanisms, it is of interest to note that the tail flick reflex in the rat is enhanced after spinal cord transection (Irwin et al. 1951). Furthermore, there is good reason to believe that the flexion reflex and pain sensation in humans are under descending control (e.g., McGrath et al. 1981; Willer et al. 1979).

A systematic investigation of the descending inhibitory system controlling the flexion reflex was begun by R. M. Eccles and Lundberg (1958; 1959b) and continued in Lundberg's laboratory (Andén et al. 1964a,b,c, 1966a,b,c; Engberg et al. 1965, 1968a,b,c,d; Grillner et al. 1967; Holmqvist and Lundberg 1959, 1961; Jankowska et al. 1967a,b, 1968). In addition to the flexion reflex per se, Lundberg's group has investigated the descending control of pathways responsible for primary afferent depolarization, the group Ia and Ib reflexes, ascending pathways activated by the FRA, as well as the influence on spinal cord mechanisms of a number of descending pathways besides the raphe- and reticulospinal tracts. The thrust of the effort of Lundberg's group has been directed primarily toward understanding motor control rather than sensory mechanisms. As it turns out, however, much of their evidence on brain stem pathways involved in motor control is directly relevant to the recent studies on the analgesia systems. This is not surprising in light of the use of flexion reflexes as tests of analgesia.

R. M. Eccles and Lundberg (1959b) showed that high-threshold muscle afferent fibers have little effect on the size of the monosynaptic reflex of either flexor or extensor motoneurons in the decerebrate cat (Fig. 26, circles). However, the same afferent fibers have a powerful facilitatory effect on flexor motoneurons and an inhibitory action on extensor motoneurons after the spinal cord is transected (Fig. 26, triangles). A similar change could be demonstrated in the effectiveness of conditioning volleys in joint or cutaneous nerves. Intracellular recordings from motoneurons showed that large increases in the sizes of postsynaptic potentials produced by the flexion reflex afferents could be seen after spinal transection. The conclusion from these experiments was that there must be a tonic descending inhibition of the flexion reflex pathway originating from neurons located in the brain.

The tonic descending inhibitory pathways were confirmed by Holmqvist and Lundberg (1959) to affect the reflex pathways from groups Ib, II, and III muscle afferents, but not the recurrent inhibitory or the group Ia paths. The fibers that inhibit the motoneurons on one side of the cord were shown to descend bilaterally in the dorsal part of the lateral funiculi. After the reflex actions that are under tonic inhibitory control from the brain were defined, selective lesions were placed in the brain stem to locate the region containing the neurons responsible for the inhibition (Holmqvist and Lundberg 1961). For example, a low pontine lesion caused a release of inhibition of the pathways from the FRA to both extensor and

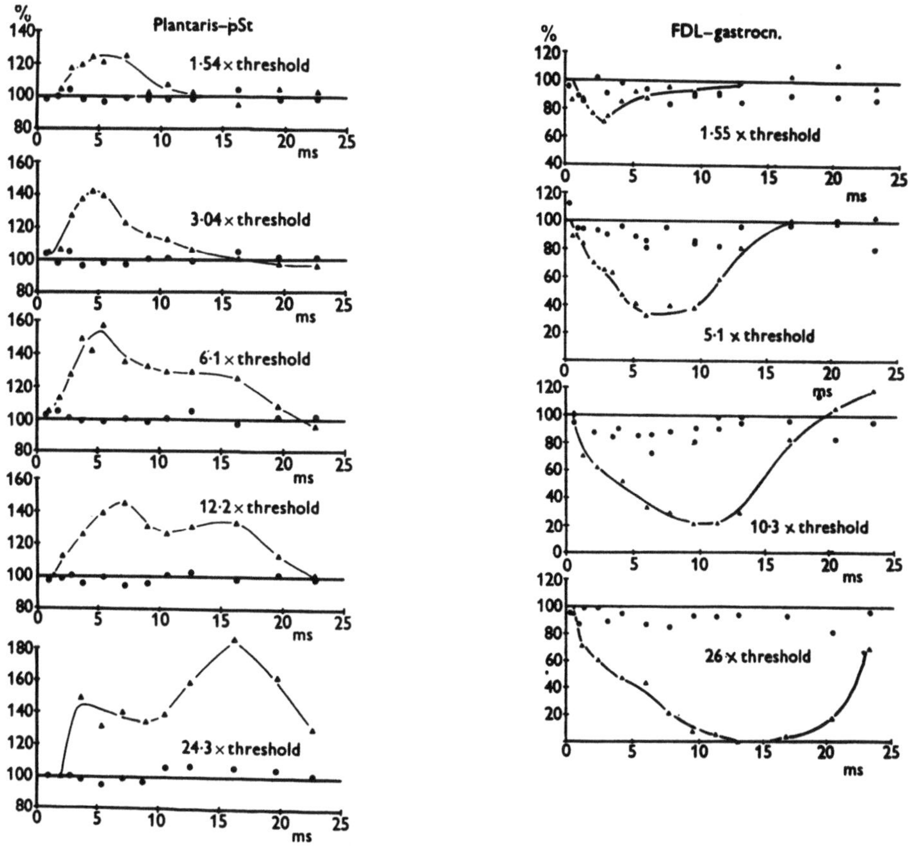

Fig. 26. Effects of volleys in high-threshold muscle afferent fibers upon flexor and extensor monosynaptic reflexes in decerebrate and spinalized states. *Left*, the effects of conditioning volleys in afferent fibers from the plantaris muscle on the monosynaptic reflex produced by stimulation of the posterior biceps-semitendinosus (*PBSt*) nerve. The size of the monosynaptic reflex is indicated as percent of the unconditioned control size. For each graph, a different strength of stimulation was used for stimulating the plantaris nerve, ranging from 1.53 to 24.3 times the threshold for the largest afferent fiber. Little change was produced in the reflex when the animal was decerebrate (*filled circles*), but large increases resulted at a variety of conditioning-testing intervals after spinalization (*triangles*). *Right*, similar changes were seen when conditioning volleys were in the flexor digitorum longus (*FDL*) nerve and the gastrocnemius-soleus monosynaptic reflex was examined, except in this case the reflex was inhibited; the inhibition was greatly enhanced by spinalization. (R. M. Eccles and Lundberg 1959b)

flexor motoneurons. The conclusion of the lesion study was that inhibitory neurons controlling both flexor and extensor motoneurons are in the medial part of the medulla. The excitatory pathway from the FRA to flexor motoneurons was only released after a lesion in the medulla, indicating a separate descending inhibitory control of this reflex pathway.

In later experiments, lesions were placed in the raphe nuclei of the pons and medulla (Engberg et al. 1968b). The lesions were nearly complete and resulted

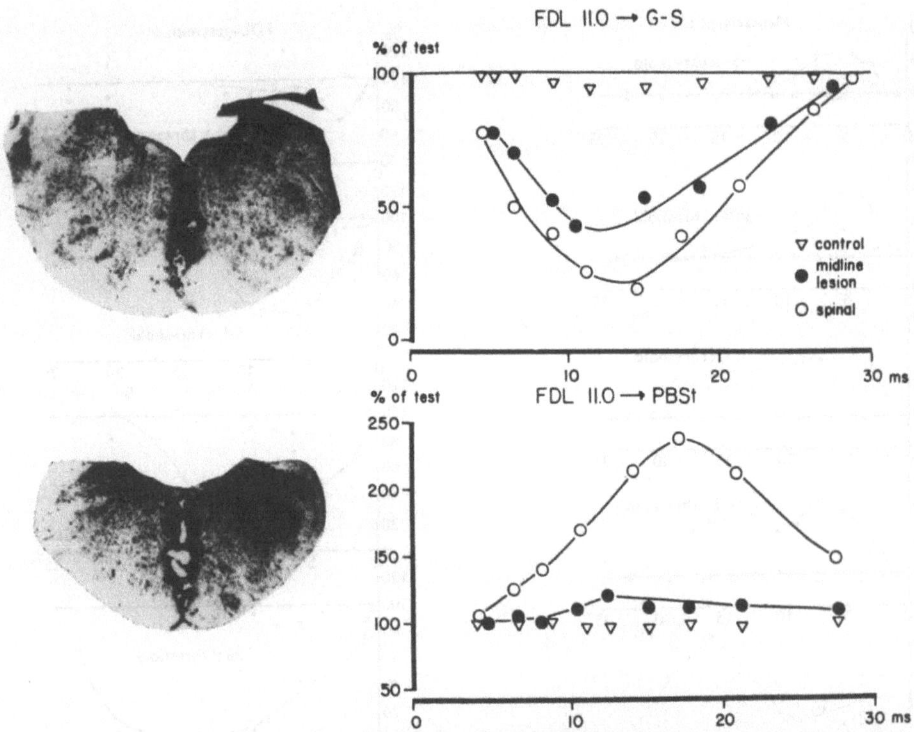

Fig. 27. Effect of raphe lesion on transmission in the flexion reflex pathway. The animal was decerebrate. Before the raphe lesion, conditioning volleys in the flexor digitorum longus (*FDL*) nerve had little effect on the monosynaptic reflexes produced by stimulation of the gastrocnemius-soleus (*G-S*) or posterior biceps-semitendinosus (*PBSt*) nerves (*triangles*). After an extensive lesion of the raphe nuclei, the G-S reflex was inhibited and the PBSt reflex slightly facilitated (*filled circles*). After spinal transection, the inhibition was increased further and the facilitation was markedly enhanced (*open circles*). (Engberg et al. 1968b)

in a partial removal of the tonic descending inhibition of the flexion reflex pathways that had previously been present in the decerebrate state (Fig. 27). The most prominent inhibition that remained after the raphe lesion was of the excitatory pathway from the FRA to flexor motoneurons.

Stimulation in the medial part of the brain stem reticular formation, including the NRM and the magnocellular part of the nucleus gigantocellularis (Fig. 28; Engberg et al. 1968c), resulted in an inhibition of the flexion reflex pathway. The inhibition was produced by stimuli that were adjusted so as not to result in inhibitory postsynaptic potentials in motoneurons or in primary afferent depolarization. Comparable stimuli resulted in postsynaptic inhibition of interneurons (Fig. 29; Engberg et al. 1968d), and so the decreased transmission in the flexion reflex path could be attributed to an action at the level of the interneurons (Engberg et al. 1968c,d). The descending inhibitory pathway was in the dorsal part of the lateral funiculus and was named the "dorsal reticulospinal system" (Engberg et al. 1968c). The conduction velocity of the fastest axons in the pathways was at least 20 m/s.

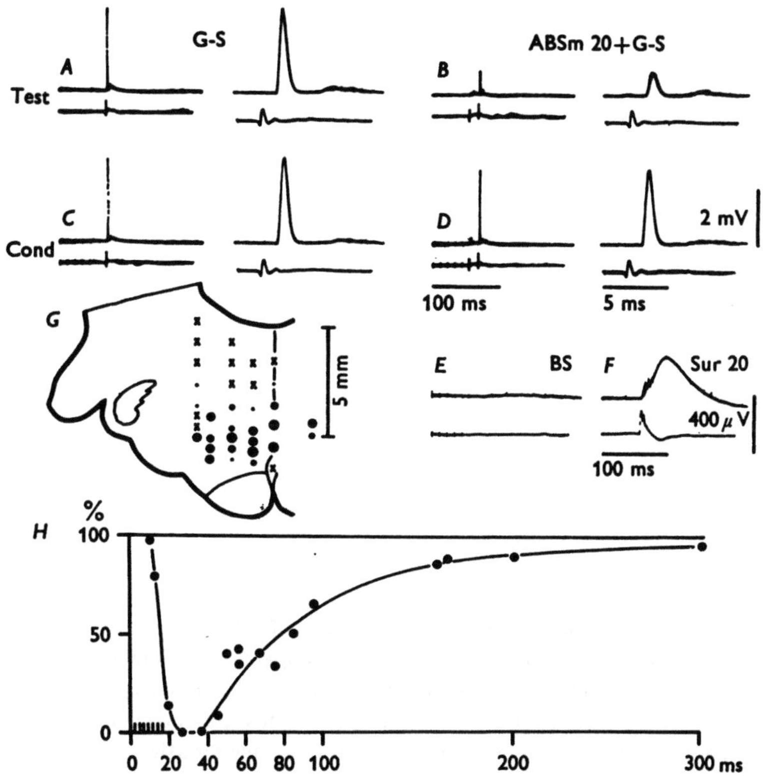

Fig. 28A–H. Inhibition of activity in the flexion reflex pathway by stimulation in the medial brain stem. In **A** is shown the monosynaptic reflex produced by stimulation of the gastrocnemius-soleus (*G-S*) nerve; the reflex is seen on both a slow and a fact sweep. In **B**, the reflex is inhibited by conditioning stimulation of high-threshold afferents in the anterior biceps-semimembranosus (*ABSm*) nerve (note that the ipsilateral dorsal quadrant and the ventral quadrants of the spinal cord were cut, releasing some of the tonic descending inhibition normally found in decerebrate animals). In **C**, stimulation of the medial brain stem (*BS*) is seen not to affect the G-S reflex, whereas in **D** the same brain stem stimulation prevents the inhibitory action of the ABSm volley. **E** shows that brain stem stimulation did not evoke a dorsal root potential, whereas **F** shows that a volley in the sural nerve (*Sur*) did. **G** shows points in the brain stem that were effective in inhibiting FRA pathways (*filled circles*) or not effective (*crosses*). **H** is the time course of the inhibition of the G-S monosynaptic reflex by high-threshold ABSm muscle afferents. (Engberg et al. 1968c).

It should be emphasized that in the experiments of Engberg et al. (1968c,d), the inhibitory effects of stimulation in the medial brain stem were limited to interneurons because of the experimental preparation used (the ventral and ipsilateral dorsal quadrants of the cord were cut; the stimulus intensity was adjusted to low levels). Stimulation in the same area of the brain stem can produce inhibition or facilitation followed by inhibition of motoneurons, largely through pathways descending through the ventral part of the cord (Clineschmidt and Anderson 1970; Jankowska et al. 1968; Llinas and Terzuolo 1964; Magoun and Rhines 1946; Proudfit and Anderson 1973), as well as primary afferent depolarization

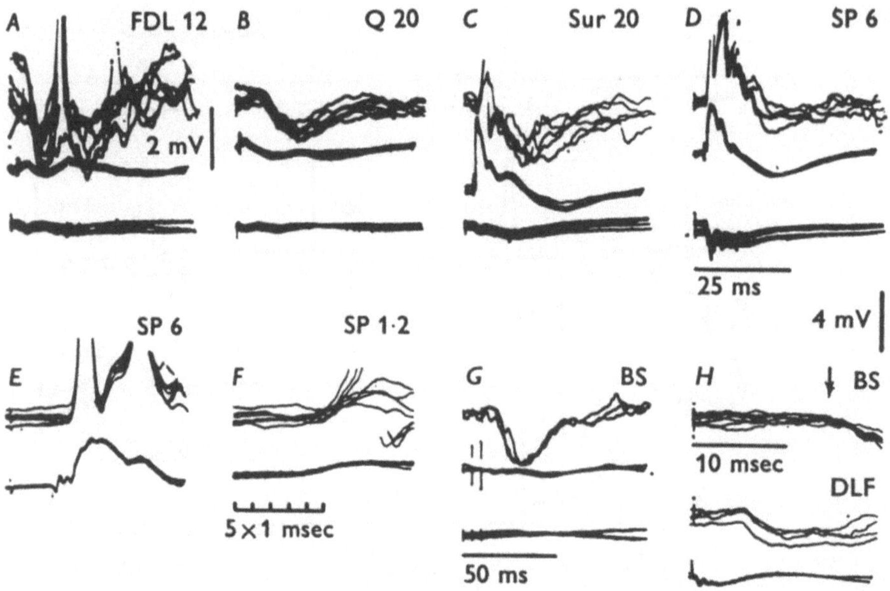

Fig. 29 A – H. Inhibitory postsynaptic potential produced in a spinal cord interneuron by stimulation in the medial brain stem. The *upper traces* show postsynaptic potentials recorded intracellularly from an interneuron located 1.8 mm deep to the surface of the spinal cord. Effects of high-threshold muscle afferents are shown in **A** and **B** (*FDL*, flexor digitorum longus; *Q*; quadriceps). Effects of cutaneous afferents are shown in **C – F** (*Sur*, sural; *SP*, superficial peroneal). Stimulus strengths are shown as multiples of threshold of largest afferent fibers. Inhibitory postsynaptic potentials produced by brief trains of stimuli in the medial brain stem (*BS*) are shown in **G** and **H**, *top* (latency indicated in H by *arrow*). A similar potential was produced by stimulation of the dorsolateral funiculus (*DLF*), **H**, *middle*. (Engberg et al. 1968d)

(Carpenter et al. 1966; Proudfit and Anderson 1974). Thus the inhibitory effect of stimulation within the medial brain stem on the flexion reflex in preparations with an intact spinal cord can be produced through a complex set of mechanisms involving inhibitory projections both to interneurons and motoneurons, as well as primary afferent depolarization (presumably resulting in presynaptic inhibition).

The brain stem pathways producing inhibition of the flexion reflex are incompletely known, but candidate pathways can be inferred from anatomical studies. Brain stem neurons whose axons descend in the dorsal part of the lateral funiculus include cells in the NRM and the magnocellular part of the nucleus reticularis gigantocellularis (Fig. 30; Basbaum et al. 1978; Basbaum and Fields 1979; Brodal et al. 1960; Leichnetz et al. 1978; Martin R. F. et al. 1978; Martin G. F. et al. 1979; Tohyama et al. 1979a,b; Watkins et al. 1980; cf. Bobillier et al. 1976; Kuypers and Maisky 1977). Cells of the medial medulla are most likely the ones forming the dorsal reticulospinal system of Lundberg. As can be seen in Fig. 30, other neurons projecting in the dorsolateral funiculus are found in the lateral pons and in the red nucleus. The reticulospinal pathways that presumably result

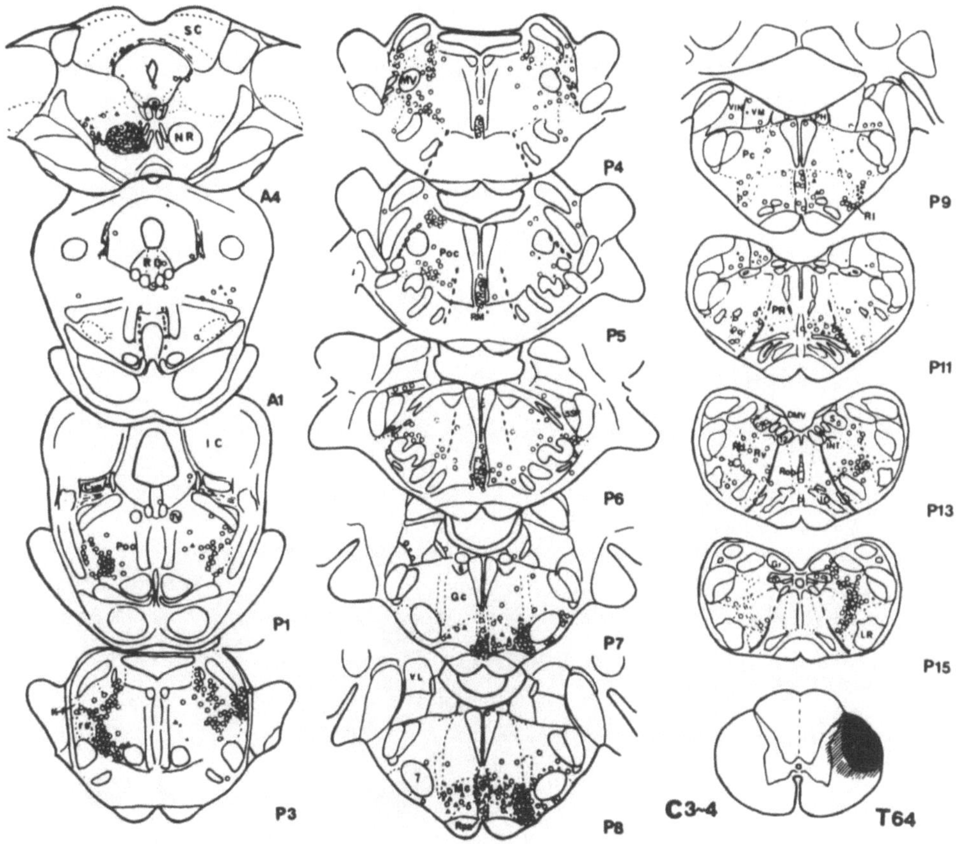

Fig. 30. Locations of brain stem neurons that project to the spinal cord through the dorsal part of the lateral funiculus. Horseradish peroxidase was injected into the cord at C3 – 4, as shown at the *bottom right*. (Tohyama et al. 1979a)

in inhibitory postsynaptic potentials in motoneurons and in primary afferent depolarization descend from the pontomedullary reticular formation and from the nucleus raphe pallidus and obscurus in the ventral part of the spinal cord (Fig. 31; Basbaum and Fields 1979; Basbaum et al. 1978; Kuypers and Maisky 1977; Nyberg-Hansen 1965; Petras 1967; Tohyama et al. 1979a). Other important cell groups whose axons travel in the ventral part of the spinal cord white matter include the catecholamine-containing neurons of the locus ceruleus, subceruleus, and dorsolateral pontine tegmentum (Fig. 31; Westlund et al. 1981).

A series of pharmacological studies on the flexion reflex pathways has been done, chiefly by Lundberg's group and by Anderson and his associates. The initial assumption in these experiments was that all of the monoamines in the spinal cord are contained in axons and synapses in the spinal cord originating from neurons whose cell bodies are in the brain stem (Andén et al. 1964a; Anderson and Holgerson 1966; Carlsson et al. 1964; Dahlström and Fuxe 1965). Thus actions of drugs affecting monoaminergic transmission are presumed to do so by way of changes in the turnover of transmitter in the terminals of pathways

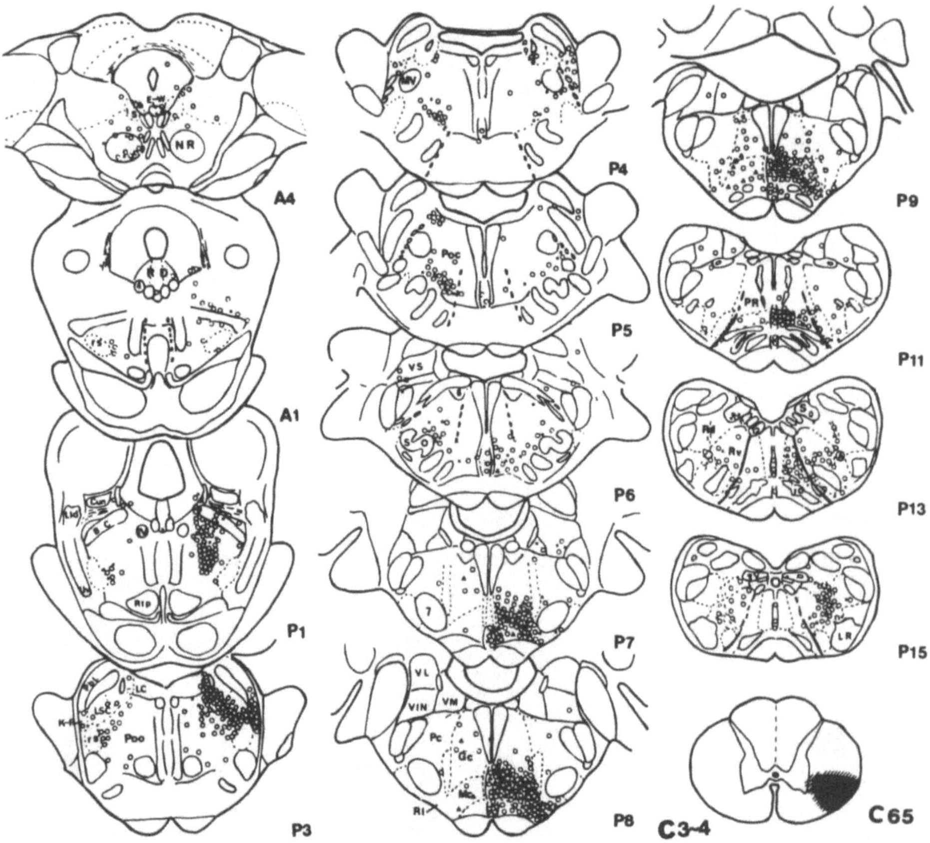

Fig 31. Locations of brain stem neurons that project to the spinal cord through the ventral part of the lateral funiculus. Horseradish peroxidase was injected into the cord white matter at C3 – 4, as shown at the *bottom right*. (Tohyama et al. 1979a)

descending from the brain stem. However, this assumption may not be completely justified, at least for the serotonin system, since there is increasing evidence for the presence of intrinsic serotonin-containing neurons in the spinal cord (Clineschmidt and Anderson 1970; Clineschmidt et al. 1971; Johns et al. 1981; Ritchie and Leonard (to be published), Shibuya and Anderson 1968).

The Lundberg group found that administration of L-dopa, the precursor of norepinephrine and dopamine, resulted in a marked enhancement of the flexion reflex evoked by pinching the skin (Andén et al. 1964c). The enhanced reflex was shown to be due to a delayed discharge of motoneurons occurring at a latency of some 200 ms after the stimulus. There was no increase in the early component of the flexion reflex, and in fact this part of the reflex could be depressed. A similar enhancement of the late flexion reflex was produced by the administration of 5-HTP, the precursor of serotonin.

Baker and Anderson (1970) showed that L-dopa increases the monosynaptic reflex. Lower doses might increase the short latency component of the flexion reflex or leave it unchanged, but higher doses reduce it. Anderson and Shibuya

(1966) found a reduction in the short latency component of the flexion reflex following 5-HTP administration. However, the monosynaptic reflex is considerably enhanced by 5-HTP (Anderson and Shibuya 1966). Motoneuronal excitability is increased by 5-HTP, since it produces an increased background discharge of motoneurons (Andén et al. 1964b; Anderson and Shibuya 1966). Interestingly, these changes produced by L-dopa and 5-HTP are also seen after chronic transection of the spinal cord, which eliminates the descending monoaminergic pathways (Baker and Anderson 1970; Shibuya and Anderson 1968). This is consistent with the idea that the pharmacological effects of 5-HTP are partly due to actions on the serotonin-containing neurons of the spinal cord, but an explanation of the effect of L-dopa is needed.

Complex interactions between monoamine synapses in the spinal cord and the flexion reflex pathway have been demonstrated in experiments using *p*-methoxyphenylethylamine (PMPEA), a phenylethylamine derivative that enhances the activity at both noradrenergic and serotonergic synapses, perhaps by releasing monoamines (Ashkenazi and Haber 1975; Walker et al. 1970). Systemic injections of PMPEA enhance the monosynaptic reflexes of flexor and extensor motoneurons, but can depress or enhance the early component of the flexion reflex. Intracellular recordings reveal that motoneurons are depolarized by systemically administered PMPEA, accounting for the enhanced monosynaptic reflexes (Jordan and Willis 1972). Interneuronal activity that may contribute to the flexion reflex response is decreased by PMPEA (Jordan and Willis 1973). However, the flexion reflex may be decreased or increased, depending upon whether the reduction in the polysynaptic excitatory postsynaptic potential is more or less significant than the depolarization of the motoneurons (Jordan and Willis 1972). Microiontophoretic application of PMPEA in the vicinity of interneurons or motoneurons causes a depression in their activity, suggesting that the depolarization of motoneurons following systemic administration of PMPEA is due to a disinhibition (Jordan and McCrea 1976). A similar depression is produced by microiontophoresis of norepinephrine or serotonin (Jordan and McCrea 1976). One implication of these experiments is that activation of descending monoaminergic pathways could have quite different results, depending on the spinal cord targets of the particular pathway involved.

Analyses of the actions of L-dopa on the spinal cord by the Lundberg group revealed the following. L-Dopa depresses transmission from the FRA to motoneurons, primary afferent terminals, and ascending pathways in a manner that resembles the depression observed in decerebrate animals (Andén et al. 1966b). Accompanying the depression in the short-latency component of the flexion reflex was the appearance of a large, long-latency flexion reflex that could be attributed to long-lasting activity within the spinal cord. These actions of L-dopa could be prevented by inhibition of dopa decarboxylase or by pretreatment with reserpine, reversed by blockers of α-receptors but not β-receptors, and enhanced by inhibition of monoamine oxidase; chronic spinalization reduced but did not eliminate the effects of L-dopa (Andén et al. 1966a). Complex actions of L-dopa were described on a number of spinal cord pathways, including appearance of primary afferent depolarization (PAD) of group Ia afferents by FRA volleys and also inhibition by the FRA of the PAD in Ia fibers produced by FRA volleys (Andén et al. 1966c). A similar effect on contralateral group Ia fibers was also seen (Jankowska et al. 1966). The late flexion reflex was shown to have a reciprocal organization (Jankowska et al. 1967a). Interneurons that may be responsible for the late flexion reflex are located in the lateral part of lamina VII (Jankowska et al. 1967b). Finally, L-dopa was shown to allow the FRA to activate static gamma motoneurons (Grillner et al. 1967).

The hypothesis that these actions of L-dopa are due to an enhanced synthesis and release of norepinephrine from descending noradrenergic fibers was tested by administration of a norepinephrine-releasing agent (4,α-dimethyl-metatyramine) in cats that had an acute spinal transection (Fedina et al. 1971). The release of norepinephrine was accompanied by depression of the short-latency component of transmission from the FRA to motoneurons, primary afferent terminals, and ascending pathways, but the late flexion reflex did not appear. The action of the norepinephrine-releasing agent was prevented by administration of α-adrenergic blockers and by pretreatment with reserpine or a blocker of tyrosine hydroxylase. Although it was proposed that the depression of the short-latency component of the flexion reflex was due to release of norepinephrine from descending noradrenergic fibers, it was still possible to observe the same action in animals that had chronic spinal cord transection, which presumably had caused the complete degeneration of the long descending monoamine pathways.

The involvement of monoamines in the control of the flexion reflex by descending pathways was tested further by administration of reserpine (Engberg et al. 1968a). This agent is known to release monoamines, including norepinephrine and serotonin, from central monoamine synaptic terminals. Reserpine depressed transmission from the FRA to motoneurons and to primary afferents in a manner similar to that produced by administration of L-dopa or 5-HTP, and a late flexion reflex appeared. However, reserpine had no effect on transmission from the FRA after chronic spinal cord transection. Since the effect of reserpine was partially reversed by a serotonin antagonist and not by an α-blocker, it was proposed that reserpine acts by release of serotonin from descending fibers.

These observations on the action of reserpine led to a study of the effects of drugs acting at monoamine synapses on the tonic inhibition of the flexion reflex pathways seen in decerebrate preparations (Engberg et al. 1968b). The tonic descending inhibition was partially blocked by serotonin antagonists but not reliably by an α-adrenergic antagonist. A monoamine oxidase inhibitor increased the level of the tonic descending inhibition. Since the descending serotonin pathways arise in large part from the raphe nuclei (Dahlström and Fuxe 1965), lesions of the raphe nuclei were made to see if the tonic descending inhibition could be interrupted. The lesions substantially reduced but did not completely eliminate the tonic descending inhibition, suggesting that this may derive in part from the raphe nuclei but that other pathways as well may be involved. Stimulation in the NRM has recently been shown to enhance the synthesis of serotonin in the spinal cord (Bourgoin et al. 1980).

Although pharmacological evidence and the effects of lesions of the raphe nuclei support the idea that serotonergic pathways contribute to the inhibition of flexion reflex pathways, there are several observations that lead to the hypothesis that other, nonserotonergic pathways are important as well (Clineschmidt and Anderson 1970; Engberg et al. 1968b). These include the findings that: (a) the inhibition may not be completely blocked by serotonin antagonists; (b) some inhibition remains after virtually complete destruction of the raphe nuclei by lesions; and (c) the conduction velocity of the fastest-descending fibers producing inhibition is too fast for these fibers to belong to serotonergic neurons. However, none of these arguments is conclusive. Serotonin blockers may not be able to produce a complete antagonism of serotonergic transmission, especially inhibitory transmission, in the CNS, since the receptors involved appear to be different from peripheral serotonin receptors (Haigler and Aghajanian 1974; Peroutka et al. 1981; Peroutka and Snyder 1979). Even complete lesions of the raphe nuclei would not eliminate all of the serotonin-containing neurons that project to the spinal cord, since many of these cells are located in the reticular formation well lateral to the raphe (Bowker et al., in press; Watkins et al. 1980). Finally, the conduction velocities of the descending serotonergic axons have not been determined directly. The objection that axons conducting at 10 – 20 m/s could not be serotonergic is based on the observation by Dahlström and Fuxe (1965)

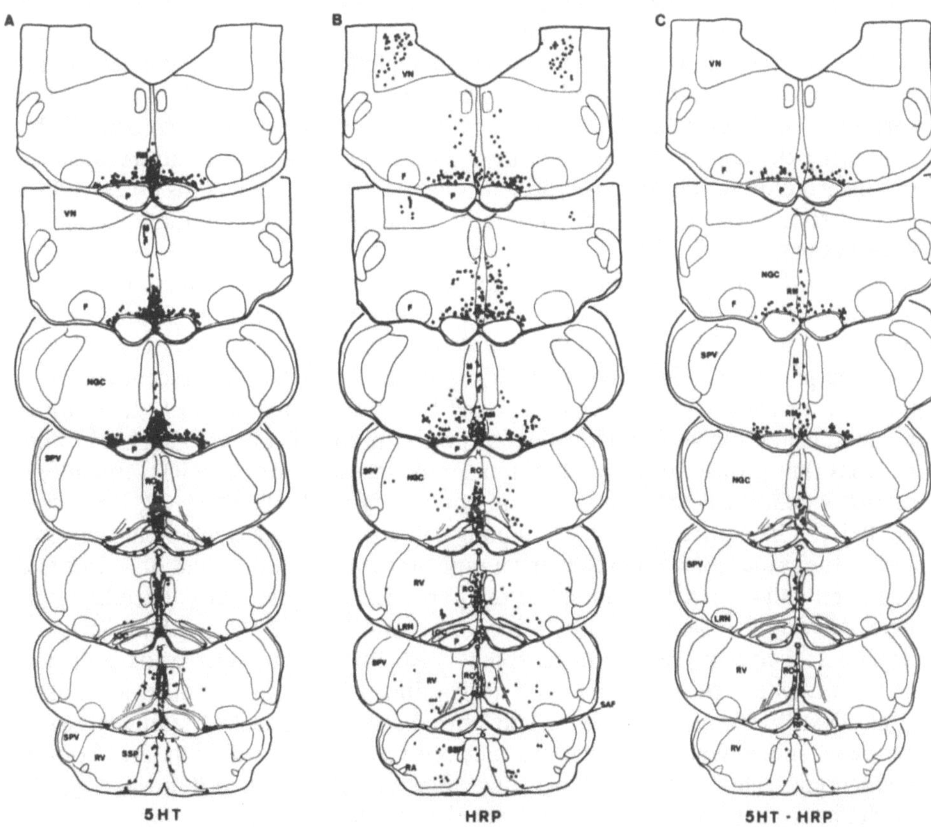

Fig. 32 A – C. Locations of serotonin-containing neurons in the brain stem that project to the spinal cord of the rat. The cells plotted in **A** were shown to contain serotonin by immunohistochemistry (*5-HT*, 5-hydroxytryptamine). The cells in **B** were labelled by horseradish peroxidase (*HRP*) injected into the lumbar spinal cord, and so these cells project to the cord. In **C** are shown cells that were doubly labeled: these are serotonin-containing cells that project to the lumbar cord. (Bowker et al., to be published)

that the serotonin-containing axons seen by fluorescence histochemistry in the rat spinal cord are not greater than 1 – 2 μm in diameter and do not appear in the light microscope to be surrounded by a myelin sheath. However, this evidence from light microscopy is not compelling, especially since it is known that in the rat spinal cord there are many myelinated fibers with diameters less than 1 μm when viewed in the electron microscope (Matthews 1968). Measurements of the conduction velocities of raphe-spinal axons reveal a range of 1 – 67 m/s; axons originating from neurons in the NRM tend to conduct at faster rates than do axons from cells of the raphe pallidus (West and Wolstencroft 1977). It was suggested that perhaps there is a higher percentage of serotonergic raphe-spinal neurons in the raphe pallidus nucleus than in the NRM, but recent work using a double-label technique demonstrates that a very high percentage of neurons in the NRM both project to the spinal cord and contain serotonin (Fig. 32; Bowker et al., in press). However, some very large neurons in the NRM that project to the spinal cord do not stain for serotonin (Bowker, Personal communication); these neurons could well have large axons and thus account for the population of cells in that nucleus having very high conduction velocities. Since few cells in the NRM have very slow axons, it can be concluded that many of the serotonin-containing axons from this nucleus are myelinated. A recent study attempted to sort out which groups of axons from the raphe nuclei are serotonergic by measuring the conduction velocity distribution before and after administration of a neurotoxin which destroys the serotonergic raphe-spinal axons (Wessendorf et al. 1981). It was concluded that many of the serotonin-containing axons are myelinated.

In view of the pharmacological evidence that serotonergic and catecholaminergic neurons in the brain stem may have inhibitory actions on flexion reflex pathways, it is of interest to know which nuclei are possible sources of serotonergic and catecholaminergic terminals in the spinal cord. Descending serotonergic neurons have been demonstrated by a double-labeling technique, using horseradish peroxidase (HRP) injections into the spinal cord to label brain stem neurons that project to the cord by retrograde transport of the HRP, and an immunohistochemical stain for serotonin-containing cell bodies (Fig. 32; Bowker et al. 1981). The main serotonin-containing cell groups in the brain stem that project to the spinal cord are in the nuclei raphe magnus, pallidus, and obscurus and in the reticular formation dorsal to the pyramid. The descending noradrenergic neurons can be demonstrated by injections of antibody to the synthetic enzyme, dopamine-β-hydroxylase, into the spinal cord, followed by retrograde transport of this enzyme to the brain stem and its demonstration by immunohistochemistry (Fig. 33A; Westlund et al. 1981). The labeled cell groups are in the nucleus locus ceruleus, subceruleus, and several other cell groups of the dorsolateral pontine tegmentum. This technique does not label noradrenergic cell groups in the medulla, so these cells appear not to project to the spinal cord.

Other catecholaminergic neurons with axonal terminations in the spinal cord utilize either dopamine or epinephrine. The dopamine-containing cells are located in the hypothalamus (Fig. 33B; Blessing and Chalmers 1979), and the epinephrine-containing cells are in the medulla just dorsolateral to the inferior olive and extending to the level of the facial nucleus (Fig. 33C; Ross et al. 1981).

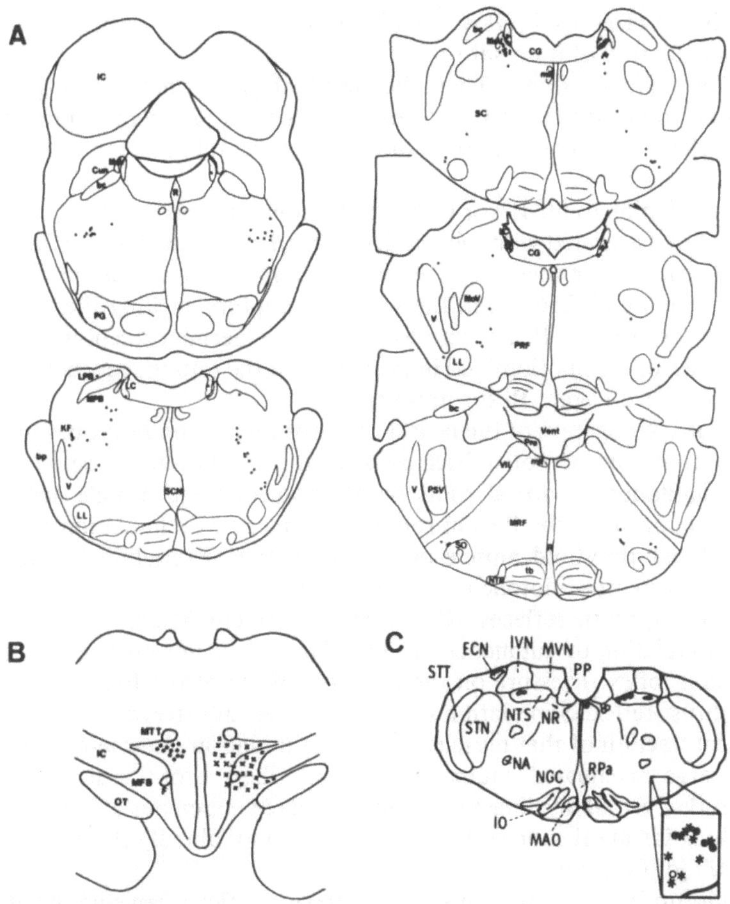

Fig. 33A – C. Catecholaminergic neurons of the brain stem that project to the spinal cord. In **A** are plotted cells labeled with antibody to dopamine-β-hydroxylase injected into the spinal cord (cervical enlargement). (Westlund et al. 1981). These cells are presumed to be noradrenergic. In **B** are shown catecholamine cells, presumed to contain dopamine, on the *left* and cells retrogradely labeled with horseradish peroxidase (injected into the thoracic cord) on the *right*. (Blessing and Chalmers 1979). In **C** are shown cells that were labeled retrogradely by a fluorescent dye, True Blue (injected into the thoracic cord, *open circles*); by an immunohistochemical stain for the synthetic enzyme, phenylethanolamine-*N*-methyltransferase, *filled circles*; or both, *asterisks*. Note the expanded view of the ventrolateral region containing the double-labeled cells. (Ross et al. 1981)

4.3 Conclusions

1. Features of the flexion reflex include: the receptive field is in skin and muscle; it may irradiate to other limbs; and it may supplant other, less powerful reflexes.
2. The flexor withdrawal reflex lasts for seconds, and its function is protective.
3. The latency of the flexion reflex can be as short as 3 ms. The latency, amplitude, and duration of the flexion reflex all depend upon stimulus intensity. A prolonged reflex must depend in part on after-discharges of central neurons,

since a prolonged reflex can still be elicited in deafferented preparations (by stimulation of a dorsal root).

4. During the flexion reflex in the hind limb, flexor muscles of the ankle, knee, and hip contract, and extensor muscles relax. If there is a crossed extensor reflex, the extensor muscles of the contralateral limb contract and the flexor muscles relax.
5. In addition to the flexor withdrawal reflex, there is another variety of flexion reflex associated with locomotion. The flexor withdrawal reflex has a higher threshold than the flexion reflex of locomotion, but takes precedence over it. There are differences in the sensory receptors eliciting these reflexes (nociceptors for the flexor withdrawal reflex; mechanoreceptors for the flexion reflex of locomotion) and in the details of the timing sequence of the contractions of different flexor muscles.
6. The investigation of the neural basis for the flexion reflex began with recordings of motoneuronal discharges from ventral roots in response to electrical stimulation of dorsal roots or of cutaneous and muscle nerves. An early component of the flexion reflex is produced by stimulation of cutaneous or of high-threshold muscle or joint afferents. Excitability changes of flexor and extensor motoneurons could also be demonstrated by conditioning monosynaptic reflexes with volleys in afferent fibers.
7. Stimulation of cutaneous afferent fibers does not invariably result in a flexion reflex. Pressure on the plantar surface of the foot causes an extensor thrust reflex. Stimulation of the skin over an extensor muscle may result in contraction of that muscle, whereas stimulation elsewhere results in flexion.
8. Large myelinated cutaneous afferent fibers evoke a short-lasting flexion reflex. Small myelinated and unmyelinated cutaneous afferent fibers evoke a long-lasting flexion reflex. The latter is undoubtedly the basis for the flexor withdrawal reflex.
9. Group II muscle afferents evoke a flexion reflex when stimulated electrically. However, it is unclear whether the group II fibers concerned arise from muscle spindles or from pressure-pain endings. Group III muscle afferents, many of which supply nociceptors, also evoke a flexion reflex, as do high-threshold joint afferents.
10. The spectrum of afferent fibers that evoke a flexion reflex has been termed the "flexion reflex afferents" (FRA). Acceptance of this term has not been universal.
11. The excitation or inhibition of motoneurons during the flexion reflex is associated with polysynaptic excitatory and inhibitory postsynaptic potentials. Presynaptic inhibition is probably also involved in the inhibitory actions during the flexion reflex.
12. Some extensor motoneurons show mixtures of excitatory and inhibitory postsynaptic potentials during the flexion reflex. Motoneurons supplying slow twitch motor units are more likely to be inhibited and those supplying fast twitch motor units are more likely to be excited.
13. The interneurons involved in the flexion reflex pathways have not been rigorously identified, although interneurons having properties appropriate to their participation in the flexion reflex have been studied on numerous occasions. Of particular interest with reference to the flexor withdrawal reflex are

nociceptive dorsal horn interneurons. These are especially abundant in the uppermost layers of the dorsal horn (laminae I and II), and they are also found in laminae V and VI.

14. Some nociceptive dorsal horn neurons receive excitation from both mechanoreceptors and nociceptors (the whole range of FRA). These interneurons are often called "wide dynamic range" cells. Many wide dynamic range interneurons are excited not only by cutaneous afferents but also by muscle and/or visceral afferents.
15. Other dorsal horn neurons are responsive only to noxious stimuli. These may be termed "nociceptive-specific" or "high-threshold" cells.
16. Interneurons in the group Ia or Ib pathways also respond polysynaptically to volleys in the FRA. These interneurons could play a role in several different reflex pathways.
17. Little is known about the flexion reflex pathways that may be concerned with the tail flick reflex.
18. The jaw-opening reflex is well studied. It has two components, one evoked by mechanoreceptors and another evoked by nociceptors. Thus, like the hind limb flexion reflex, there is a jaw-opening reflex involved in a rhythmic response (chewing), and one serving a protective function.
19. The flexion reflex can be shown to be under descending control by transecting the spinal cord. This procedure enhances the reflex.
20. Other reflex pathways, such as the group Ia and the recurrent inhibitory pathways, are not subject to tonic descending inhibition, although the group Ib pathway is.
21. Lesions of the brain stem show that the neurons responsible for the tonic descending inhibition of the flexion reflex pathway are in the medial medulla. The raphe nuclei form one component of the neural apparatus involved. The adjacent reticular formation is involved as well. The descending pathways travel in the dorsal part of the lateral funiculus and have been termed the "dorsal reticulospinal system."
22. At least part of the tonic descending inhibition of the flexion reflex pathway is mediated by inhibition of the interneurons of the pathway. There are also parallel inhibitory actions on motoneurons, via the ventral part of the cord, and presynaptic inhibitory actions.
23. Pharmacological studies using monoamine precursors, monoamine releasing agents, and monoamine receptor blockers lead to the conclusion that monoamines, especially serotonin, are involved in tonic descending inhibition of the flexion reflex. However, other pathways not utilizing monoamine transmitters may also be involved.
24. The most likely serotonin-containing neurons to participate in descending inhibition of the flexion reflex are those in the nucleus raphe magnus, since this nucleus projects to the cord in the dorsal part of the lateral funiculus and since most of the raphe-spinal neurons of this nucleus contain serotonin.
25. Other fibers descending in the dorsolateral funiculus originate in the magnocellular part of the nucleus reticularis gigantocellularis.
26. Ventrally coursing axons that might directly affect motoneurons arise from neurons in much of the medial reticular formation of the caudal brain stem, the raphe pallidus and obscurus nuclei, and the dorsolateral pons.

27. The descending fibers from the nucleus raphe magnus terminate in laminae I, II, and V – VII of the spinal cord gray matter. These fibers would presumably be particularly likely to affect interneuronal activity.
28. Catecholaminergic fibers descending to spinal cord levels originate largely from the locus ceruleus, nucleus subceruleus, and dorsolateral pontine tegmentum. However, some epinephrine-containing projections may arise in the medulla, and some dopaminergic fibers in the hypothalamus.

5 Descending Control of Spinal Cord Nociceptive Neurons

5.1 Dorsal Horn Interneurons

Afferent input directly to the dorsal column nuclei by way of primary afferent fibers does not appear to include nociceptive input (reviewed in Willis and Coggeshall 1978). All of the nociceptive information to be transmitted by sensory pathways to the brain relays in the spinal cord gray matter near the dorsal root through which that input is conveyed. Many of the ascending sensory tracts originate in the dorsal horn, and so investigations of spinal cord sensory transmission related to nociception have focused on recordings from interneurons of the dorsal horn. However, nociceptive signals that trigger the flexion reflex must also be transmitted by way of interneurons, presumably of the dorsal horn, and it is impossible to know from recordings of interneuronal activity whether a given interneuron is dedicated to sensory or motor events, or both. The close correlation between pain sensation and flexion reflexes in man (Willer 1977) make a combined sensory and motor role for nociceptive interneurons a viable hypothesis, but there is little hard evidence for or against such a possibility. Furthermore, the flexion reflex and pain sensation can be dissociated (McGrath et al. 1981; Willer et al. 1977), and so the mechanisms must be at least in part distinct. An alternative approach to the study of sensory and motor mechanisms in isolation is to examine the behavior of known sensory neurons and of motoneurons. Nevertheless, a considerable amount of useful information has been accumulated concerning the responses of dorsal horn interneurons to noxious, as well as to innocuous, inputs, and the ability of supraspinal pathways to affect the responses to these signals.

Recordings from dorsal horn interneurons that are specifically nociceptive were first made by Kolmodin and Skoglund (1960). However, many later studies failed to include such interneurons, presumably because of sampling error. Instead, it was emphasized that a large number of dorsal horn interneurons respond to innocuous mechanical stimulation as well as to noxious stimulation, although the effects of noxious stimuli are more potent (Mendell 1966; Wall 1967). Because of the apparent wide band width of mechanical signals to which many dorsal horn interneurons respond (referring to the range of intensities of the stimuli, rather than to their frequencies), these neurons have been called wide dynamic range neurons. In addition to high-threshold and wide dynamic range neurons, other neurons of the dorsal horn respond only to innocuous mechanical stimuli. These have been called low-threshold cells. Other interneurons have a predominantly proprioceptive input, and these cells can be called "deep" cells, because the sensory receptors affecting them are located in subcutaneous tissue.[1]

Nociceptive-specific or high-threshold cells are abundant in lamina I (e.g., Cervero et al. 1976; Christensen and Perl 1970). Such cells have recently been

[1] For alternative classification schemes, see Cervero et al. 1976; Menétrey et al. 1977; Price and Browe 1973

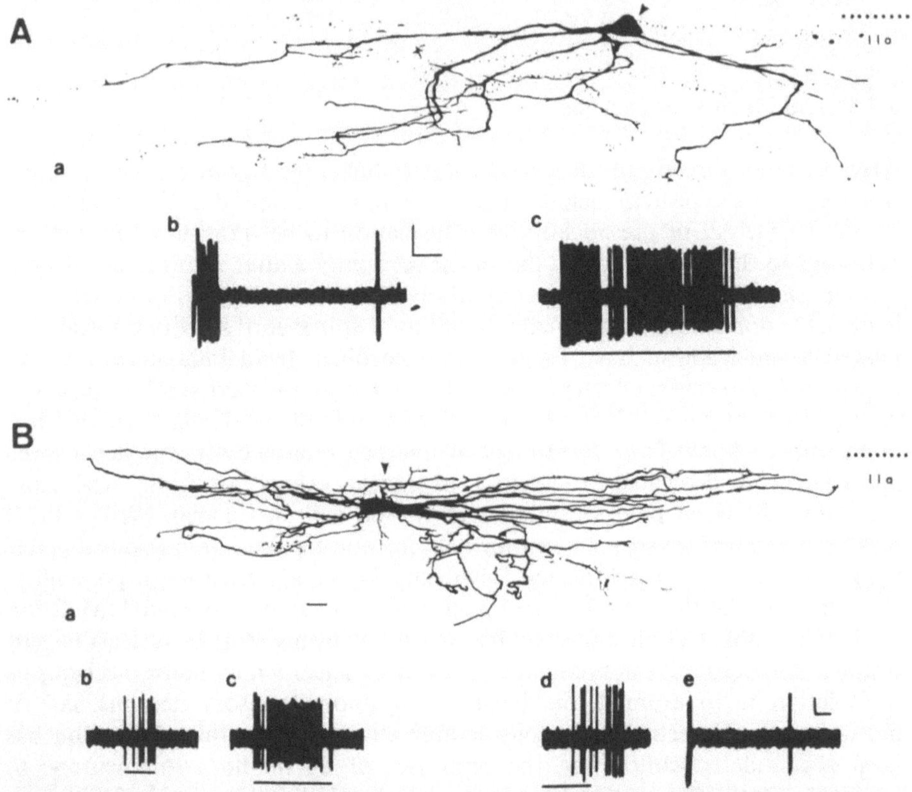

Fig. 34. The neuron in **A**, *a* is a nociceptive-specific interneuron located in the upper part of lamina II. The type is a "limitroph" (Cajal 1909) or "stalked" cell (Gobel et al. 1980). The response of the cell to electrical stimulation of the skin in *b* shows an input from small myelinated and unmyelinated fibers. The response to pinching the skin in *c* allowed the cell to be classified as nociceptive-specific (innocuous stimuli were ineffective). The neuron in **B**, *a* is a wide dynamic range interneuron located in the upper part of lamina II. The type is "central cell" (Cajal 1909) or "islet cell" (Gobel et al. 1980). The responses in *b* – *d* were, respectively, to gentle brushing, pinching, and noxious heating of the receptive field. The discharges in *e* indicate an input from both small myelinated and unmyelinated afferents. (Bennett et al. 1980)

demonstrated in the upper part of lamina II in experiments utilizing a combination of electrophysiological identification and intracellular injection of HRP as a marker substance (Fig. 34A; Bennett et al. 1980; Light et al. 1979). Similar cells have also been demonstrated in the deeper layers of the dorsal horn (e.g., Kolmodin and Skoglund 1960; Gregor and Zimmermann 1972; Price and Browe 1973; Menétrey et al. 1977). Wide dynamic range cells can be found in these same locations, although they seem proportionately more abundant in the deeper than in the more superficial layers of the dorsal horn (Fig. 34B; Bennett et al. 1980; Cervero et al. 1976; Besson et al. 1972; Gregor and Zimmermann 1972; Menétrey et al. 1977; Wall 1967). Low-threshold cells have now been described in the deep part of lamina II, as well in laminae III and IV (Light et al. 1979; cf. Menétrey et al. 1977; Wall 1967; Wall et al. 1979). Deep cells are most commonly found in the deepest part of the dorsal horn (Fields et al. 1977a; Menétrey et al. 1977; Wall 1967).

This classification scheme for dorsal horn interneurons has both the virtue and the disadvantage of simplicity. A virtue is the fact that the cells can be rapidly classified, using relatively simple screening tests. Disadvantages include the fact that the scheme is based on input from somatic mechanoreceptors, overlooking many potentially important sources of sensory input, such as thermal receptors and visceral receptors. However, when attempts have been made to specify the particular sensory receptor organs that project onto a particular dorsal horn interneuron, the number of categories of interneuron can expand dramatically (Heavner and DeJong 1973; Tapper et al. 1973). Another problem with simple classification schemes is that they tend to encourage an association between the classification of a given interneuron and a particular sensory modality. For example, it is easy to assume that low-threshold cells are concerned with the sensations of touch-pressure or flutter-vibration (see Willis and Coggeshall 1978). However, unless the dorsal horn interneuron projects to the brain, a sensory role for such a neuron cannot be assumed. Instead, the cell may play a role in reflex arcs that receive inputs from sensitive mechanoreceptors. Similarly, high-threshold neurons may simply provide inputs for high-threshold reflexes.

The presence of a tonic descending inhibitory system that controls the activity of dorsal horn interneurons was first demonstrated indirectly by R.M. Eccles and Lundberg (1959b) by recording postsynaptic potentials produced through the flexion reflex pathways from motoneurons before and after spinal cord transection. Interruption of the descending control system made the postsynaptic potentials much larger in response to the same peripheral stimulation. This finding was related to the enhancement of the flexion reflex that is seen after spinal transection and interpreted to account for this reflex change (Eccles R.M. and

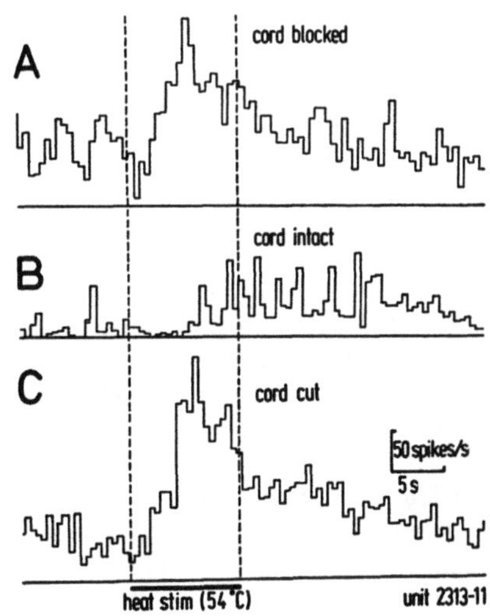

Fig. 35A – C. Effects of interruption of conduction in the spinal cord through long descending pathways upon the responses of a wide dynamic range neuron to noxious heating of the skin. In **A** the activity of the cell is shown during cold block of the cord at L1. In **B** the response to the same stimulus is shown when cord conduction was intact. In **C** is shown the response after surgical transection of the cord at L1. (Handwerker et al. 1975)

Lundberg 1959b; see Section 4). Changes in interneuronal activity following spinal transection or block have since been demonstrated by several groups (Fig. 35; Besson et al. 1975; Cervero et al. 1976, 1979; Duggan et al. 1981; Handwerker et al. 1975; Hillman and Wall 1969; Wall 1967). A direct demonstration that stimulation in the medial brain stem, in the vicinity of the NRM and the adjacent reticular formation, inhibits interneurons was provided by Engberg et al. (1968d), who recorded inhibitory postsynaptic potentials from dorsal horn interneurons following such stimulation. The descending inhibitory pathway traveled in the dorsal part of the lateral funiculus (Fig. 29).

After it was discovered that analgesia could result from stimulation in the PAG (Reynolds 1969; Section 2), it was also found that PAG stimulation causes an inhibition of dorsal horn interneurons (Fig. 36A,B; Bennett and Mayer 1979; Liebeskind et al. 1973; Oliveras et al. 1974a). It was a reasonable assumption that the inhibition of dorsal horn interneurons by PAG stimulation could explain the

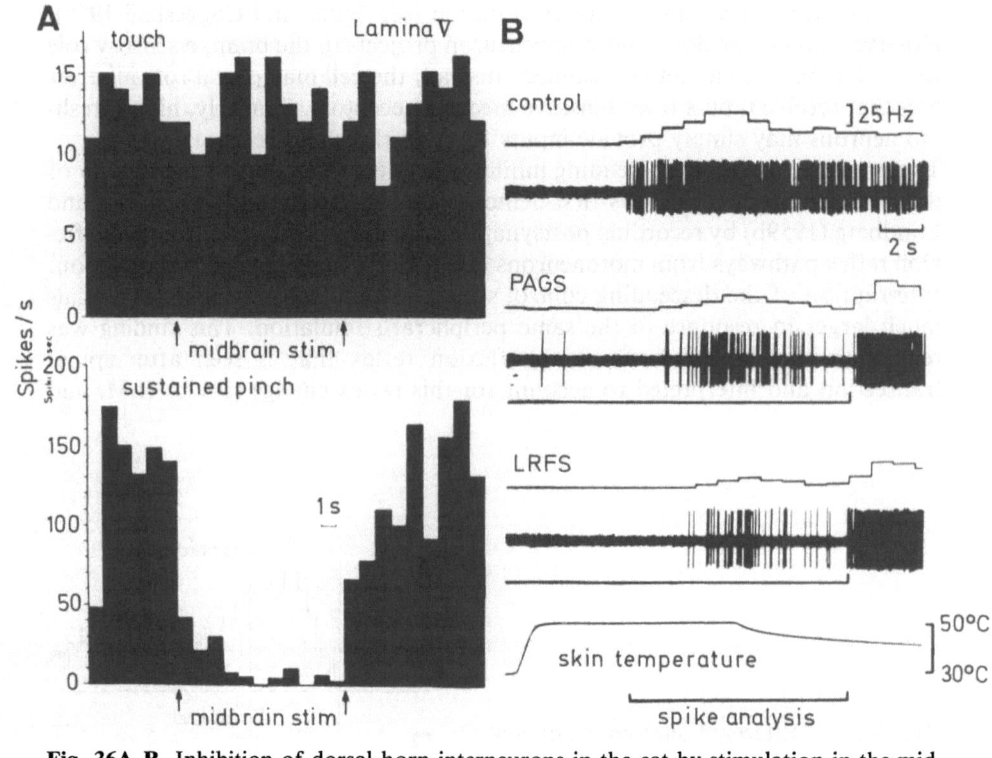

Fig. 36A, B. Inhibition of dorsal horn interneurons in the cat by stimulation in the midbrain. In **A** are shown the responses of a wide dynamic range cell to touch (*above*) and to pinch (*below*). During each response, the periaqueductal gray (*PAG*) was stimulated during the time *between the arrows*. PAG stimulation did not affect the tactile response, whereas it did inhibit the response to noxious stimulation. (Oliveras et al. 1974a). In **B** are shown the effects of midbrain stimulation on the responses of a dorsal horn interneuron to noxious heat pulses (temperature monitor at *bottom*). The control response at the *top* is shown both by a peristimulus time histogram and by the unit's activity. The other pairs of records show the inhibition produced by stimulation in the PAG and in the lateral reticular formation (*LRF*) of the midbrain. (Carstens et al. 1980a)

prolongation of the tail flick latency in stimulation-produced analgesia. Interestingly, PAG stimulation tended to inhibit the responses of wide dynamic range neurons to noxious but not to innocuous stimuli (Fig. 36A), while leaving unaffected the activity of low-threshold cells (however, see Duggan and Griersmith 1979). This selectivity of the descending inhibitory controls might help explain the observation in experiments on analgesia produced by PAG stimulation that the animals appeared to have intact responses to tactile and other sensory stimulation (Mayer et al. 1971; Oliveras et al. 1974b).

Inhibition of dorsal horn interneurons by PAG stimulation has been studied extensively by Carstens et al. (1979c, 1980a,b, 1981a). The inhibition of a given interneuron resulted from stimulation in either side of the PAG, and the effective sites could be ventral, lateral, or dorsal within the PAG. Threshold stimuli for inhibition averaged 167 μA, and inhibition increased with stimulus strength. Responses generally recovered within minutes after termination of PAG stimulation (Carstens et al. 1979c, 1980b). When stimuli were applied in the reticular formation lateral to the PAG, it was also possible to demonstrate inhibition of dorsal horn interneurons (Fig. 36B; Carstens et al. 1980a). Interestingly, when discharges were evoked by graded noxious heat pulses, the stimulus-response curves were affected differently by conditioning stimulation of the PAG and of the reticular formation (Fig. 2). Periaqueductal gray stimulation caused a reduction in the slope of the stimulus-response curve, indicating a changed gain in the sensory pathway, whereas reticular formation stimulation resulted in a shift in the curve to the right, suggesting a change in set point or threshold. Inhibition of the responses to noxious stimuli was somewhat stronger than the inhibition of the responses to innocuous mechanical stimuli when stimuli were applied in either the PAG or the adjacent reticular formation (Carstens et al. 1981a).

Besson's group found that analgesia could be produced by stimulation not only of the PAG but also of the NRM (called by them the central inferior nucleus of the raphe, after Berman 1978; Oliveras et al. 1975; see Section 2). The analgesia was soon attributed to the inhibition of dorsal horn interneurons. As mentioned earlier, this inhibition was initially demonstrated by Engberg et al. (1968d), but it has since been confirmed by a number of laboratories (Belcher et al. 1978; Duggan and Griersmith 1979; Fields et al. 1977; Guilbaud et al. 1977b; Rivot et al. 1980; Willis et al. 1977). The ability of stimulation of the NRM to produce a selective inhibition of responses to noxious, as opposed to innocuous, stimuli is controversial. However, there is a strong inhibition of the responses to electrical stimulation of small afferent fibers (Fig. 37A; Rivot et al. 1980). As in the case of the midbrain, stimulation in the reticular formation adjacent to the NRM may also result in the inhibition of dorsal horn interneurons (Fig. 37A; Rivot et al. 1980), although this action may not be as strong as the NRM inhibition. The threshold for inhibition following stimulation in the NRM is generally reported to be lower than that for PAG stimulation, often being 50 μA or less (Guilbaud et al. 1977b; Rivot et al. 1980; Willis et al. 1977). Inhibition can persist as long as 900 ms after a stimulus train lasting 100 ms (Rivot et al. 1980). Inhibition increases with stimulus frequency and train duration (Rivot et al. 1980; Willis et al. 1977).

The pathway for NRM inhibition appears to descend in the dorsal part of the lateral funiculus, since interruption of this part of the spinal cord white matter

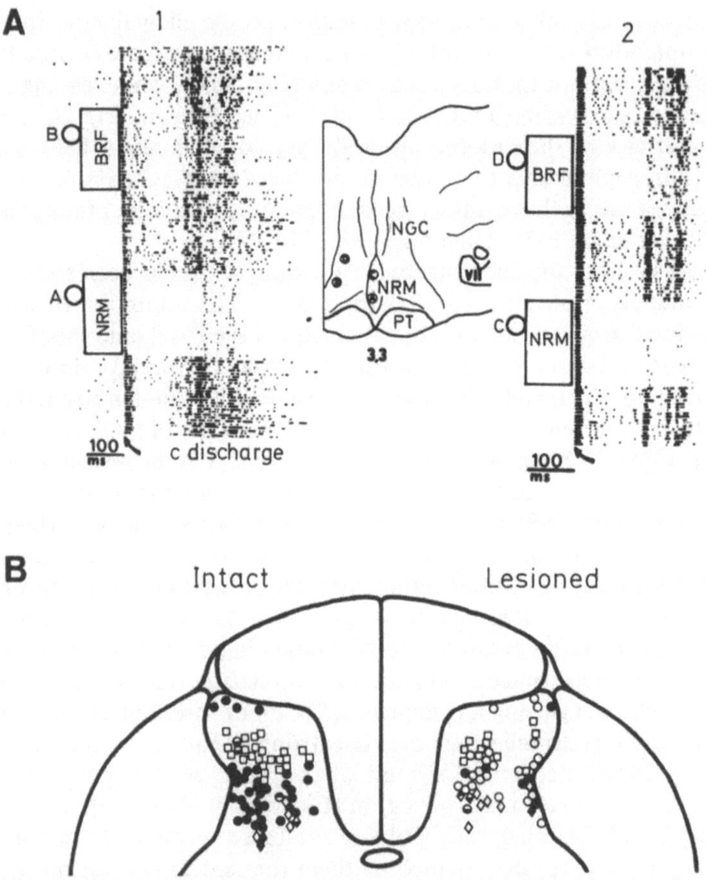

Fig. 37A, B. Inhibition of dorsal horn interneurons by stimulation in the nucleus raphe magnus (*NRM*) or adjacent reticular formation (*BRF*). In **A** are shown the effects of such stimulation on two interneurons in the dorsal horn of the rat spinal cord. Responses were to stimulation of the skin at strengths sufficient to activate A and C fibers. For cell *1*, NRM stimulation inhibited the responses, while BRF stimulation had little effect; for cell *2*, both kinds of stimulation were effective. Stimulus sites are shown in the drawing. *NGC*, nucleus gigantocellularis; *PT*, Pyramid; *VII*, facial nucleus (Rivot et al. 1980). In **B** is shown the distribution of cells in the cat that were (*filled symbols*) or were not (*open symbols*) inhibited by stimulation of NRM. For the cells on the left, the dorsolateral funiculus was intact, whereas for the cells on the right it was interrupted. *Circles*, cells having nociceptive input; *squares*, low-threshold cells; *diamonds*, cells with deep input. (Fields et al. 1977)

reduces the inhibition profoundly (Fig. 37B; Fields et al. 1977). The fastest axons of the descending pathway must be myelinated, since the latency of the inhibition caused by NRM stimulation is about 15 ms (Engberg 1968d; Guilbaud et al. 1977b; Rivot et al. 1980; Willis et al. 1977). Presumably, the inhibitory pathway from the NRM involves the raphe-spinal projection from this nucleus, which descends in the dorsolateral funiculus (Basbaum et al. 1978; Basbaum and Fields 1979; Martin R.F. et al. 1978; Martin G.F. et al. 1979; Tohyama et al. 1979a,b;

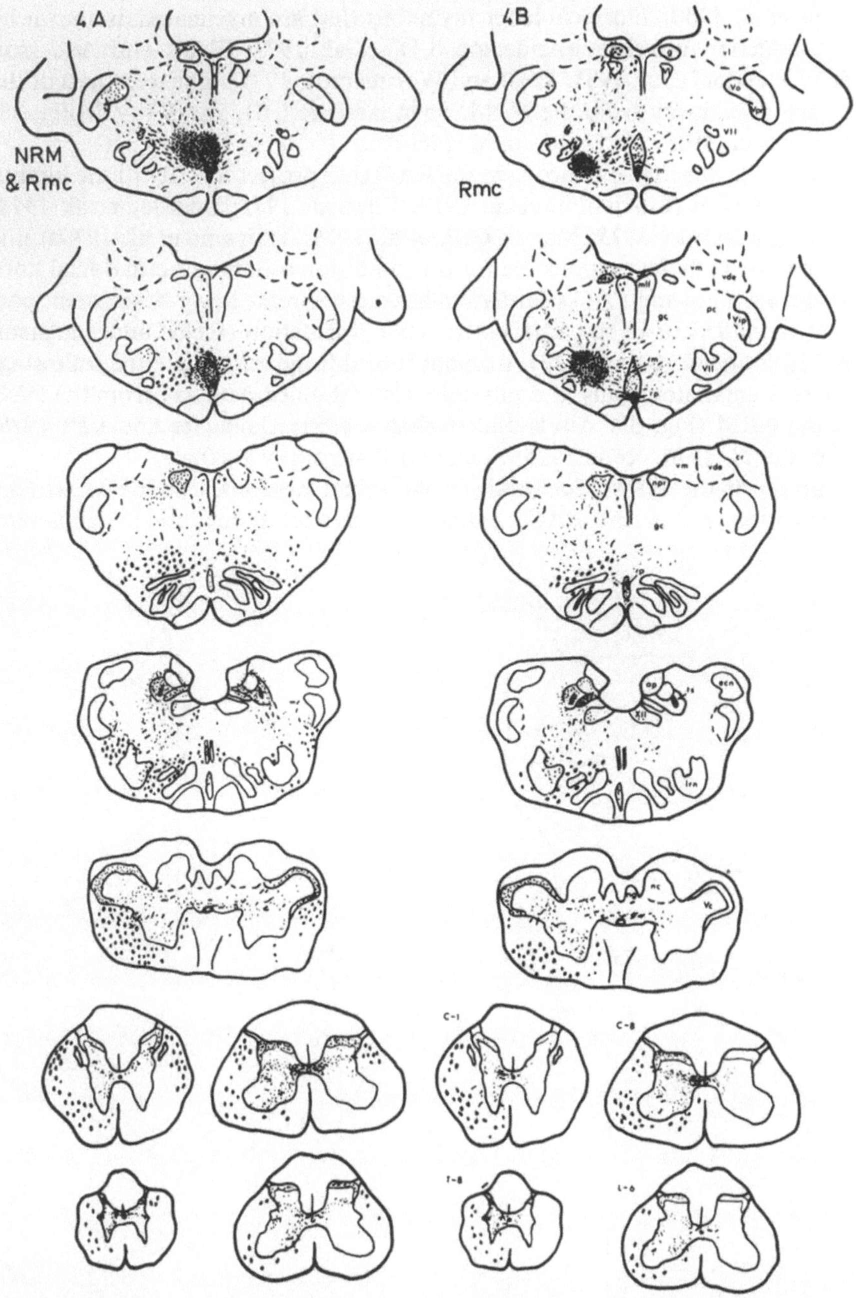

Fig. 38. Drawings showing the distribution of radiolabeled amino acid after injection into the nucleus raphe magnus (NRM) and adjacent reticular formation (*left*) or into the magnocellular part of the nucleus gigantocellularis (Nmc) (*right*). (Basbaum et al. 1978)

Watkins et al. 1980) and includes many axons that are myelinated as shown by their conduction velocities (Anderson S.D. et al. 1977; Fields and Anderson 1978; Wessendorf et al. 1981; West and Wolstencroft 1977). The terminals of the raphe-spinal pathway from the NRM are in laminae I, II, and V–VII (Fig. 38; Basbaum et al. 1978; cf. Goode et al. 1980).

Because of the paucity of neurons in the PAG that project directly to the lumbosacral spinal cord (Castiglioni et al. 1978; Edwards 1975; Kneisley et al. 1978; Kuypers and Maisky 1975; Martin G.F. et al. 1979; Tohyama et al. 1979a) it is likely that the pathway responsible for the inhibition of lumbosacral dorsal horn interneurons following PAG stimulation has a relay in the lower brain stem, possibly in the NRM or in the adjacent reticular formation (Fields and Anderson 1978; Fields and Basbaum 1978). In support of this hypothesis are the following:

1. There is an anatomically and physiologically defined pathway from the PAG to the NRM (Fig. 39; Abols and Basbaum 1981; Gallagher and Pert 1978; Pierce et al. 1976; Ruda 1975; Shah and Dostrovsky 1980).
2. Neurons in the NRM, including raphe-spinal neurons, can be excited by stimulation within the PAG, either by electrical shocks or by glutamate

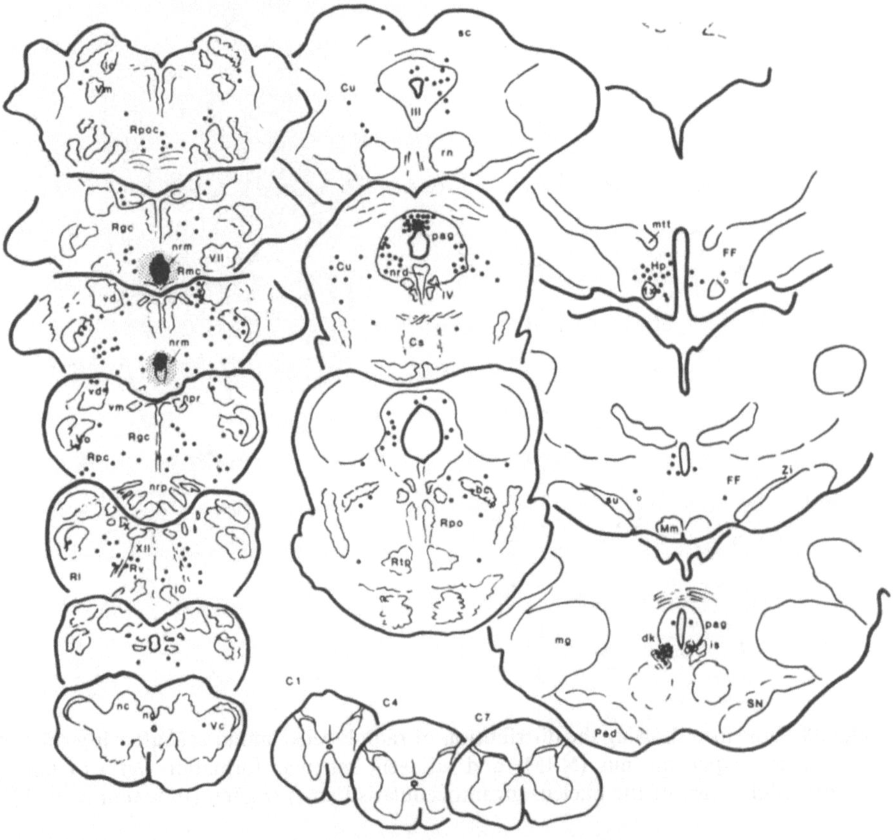

Fig. 39. Distribution of neurons labeled retrogradely by horseradish peroxidase injected into the nucleus raphe magnus (NRM). (Abols and Basbaum 1981)

injections (Behbehani and Fields 1979; Fields and Anderson 1978; Lovick et al. 1978; Oleson et al. 1978; Pomeroy and Behbehani 1979).

However, it should be stressed that there is also a synaptic link between the PAG and the reticular formation, including both the nucleus gigantocellularis and magnocellularis (Mohrland and Gebhart 1980; Pierce et al. 1976).

Dubuisson and Wall (1980) found many interneurons in the superficial layers of the dorsal horn (laminae I – III) that are excited by stimulation of the dorsolateral funiculus. Assuming that the excitation was due to activation of descending control pathways that inhibit nociceptive responses of other dorsal horn interneurons, these authors propose that the excited cells could serve as inhibitory interneurons in the pathways from the brain stem (however, cf. Cervero et al. 1979b).

Inputs to the PAG and to the NRM are under investigation by both anatomical and physiological techniques (Abols and Basbaum 1981; Anderson S.D. et al. 1977; Becker et al. 1969; Behbehani and Fields 1979; Behbehani and Pomeroy 1978; Fields and Anderson 1978; Moolenaar et al. 1976; Sakai et al. 1977; Sanders et al. 1980; Shah and Dostrovsky 1980). Neurons in the PAG can often be activated by peripheral stimulation of large, bilateral receptive fields; in some cases, no receptive field can be demonstrated, or inhibition is seen (Sanders et al. 1980; Shah and Dostrovsky 1980). Several reports indicate that the neurons of the NRM often have similar receptive field properties (Anderson S.D. et al. 1977; Behbehani and Fields 1979; Behbehani and Pomeroy 1978; Fields and Anderson 1978; Moolenaar et al. 1976). The possibility that the PAG and NRM are elements of a negative feedback system regulating nociceptive transmission has been suggested (Fig. 40; Basbaum and Fields 1978; Fields and Anderson 1978; Liebeskind et al. 1973; Sanders et al. 1980).

Recently, inhibition of spinal cord nociceptive neurons by high-intensity stimulation of widespread areas of the skin has been demonstrated; this form of inhibition has been named by Besson's group "diffuse noxious inhibitory controls" (DNIC) (LeBars et al. 1979a,b). According to the findings of Besson's group, based on experiments using rats, the DNIC mechanism has the following properties:

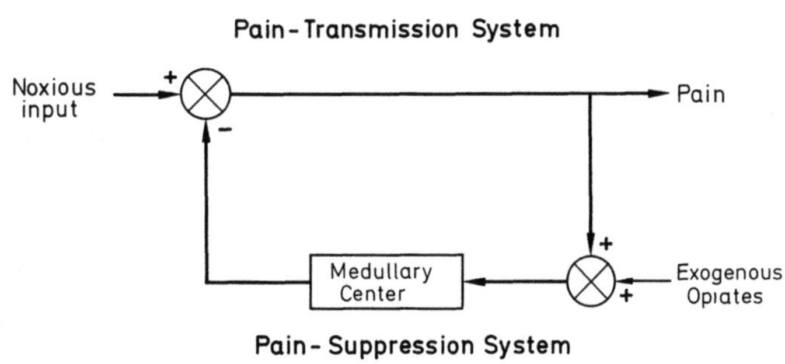

Fig. 40. Descending pathways controlling nociceptive transmission may be considered a negative feedback system. Noxious input activates an ascending pathway that causes pain sensation, but also a feedback loop operating through a control point and a relay in the medulla. The control point involves an opiate transmitter mechanism and is a site of action for exogenously administered opiates. (Basbaum and Fields 1978)

1. Inhibition requires noxious intensities of stimulation.
2. The areas of skin stimulated can be anywhere on the limbs, trunk, or face other than the excitatory receptive field.
3. The pathway involves a loop through the brain stem.
4. The main inhibitory action is on wide dynamic range neurons, but not on high-threshold, low-threshold, or deep cells. However, experiments in cats and monkeys suggest that part of the inhibitory pathway can be at the spinal level, and that high-threshold neurons may also be affected (Gray and Dostrovsky 1981; Gerhart et al. 1981b).

Besson's group proposes that the DNIC mechanism is important in the enhancement of contrast in the ascending nociceptive pathways (LeBars et a. 1979b). Wide dynamic range neurons that are not excited by a noxious stimulus in their receptive fields are instead inhibited; the distribution of excitation and inhibition in the population of wide dynamic range cells provides the crucial signal for the sensory discrimination of pain.

Several studies have addressed the pharmacology of the descending control systems acting on dorsal horn interneurons. It has been found, for instance, that microinjections of morphine into the PAG result in the inhibition of dorsal horn interneurons (Fig. 41; Bennett and Mayer 1979). It is not known what the precise action of morphine is in this brain stem structure; generally speaking, morphine depresses neural activity, and so it is possible that its action in the brain stem is to produce disinhibition by blocking tonically active inhibitory pathways onto projection neurons. Morphine microinjections into the PAG can activate (or in other cases inhibit) cells in the NRM and adjacent reticular formation (Behbehani and Fields 1979; Behbehani and Pomeroy 1978; Fields and Anderson 1978; Mohrland and Gebhart 1980; Oleson et al. 1978). Systemic morphine injections can also excite NRM cells (Anderson S.D. et al. 1977); the activity of the same neurons is depressed by naloxone.

There have been reports that naloxone decreases the inhibitory actions resulting from electrical stimulation in the NRM (Rivot et al. 1979). The responses of

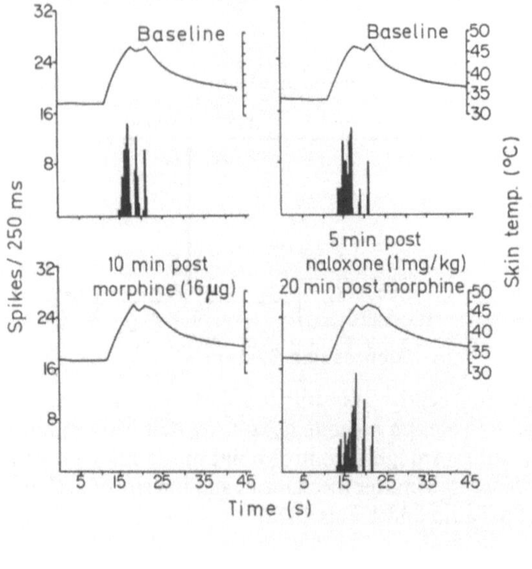

Fig. 41. Effect of morphine microinjection into the periaqueductal gray on the response of a wide dynamic range neuron to noxious heat stimuli. The time course of noxious heat pulses is shown at the *top of each panel.* The *upper two* histograms show control responses, while the *lower left* one shows the lack of response after morphine. The *lower right* panel shows that the effect of morphine could be reversed by naloxone. (Bennett and Mayer 1979)

dorsal horn interneurons to C fiber volleys are enhanced by naloxone (Rivot et al. 1979). These observations are consistent with a role for endogenous opioids in the descending inhibitory pathways. However, other investigators report negative results when they have tried the effects of naloxone on the inhibition from the PAG (Carstens et al. 1979a; Duggan and Griersmith 1979; Jurna 1980) or NRM (Duggan and Griersmith 1979). If the projection neurons are being activated directly by electrical stimuli, presumably blocking opiate synapses might have little effect; on the other hand, if projection neurons were being affected transsynaptically through a network involving opioid-containing neurons, then naloxone should have an action. Thus, it is possible that differences in the effectiveness of naloxone in altering the inhibition produced by stimulation in the PAG or NRM might depend upon the details of the experimental procedure and perhaps upon the specific anatomical arrangements in different animal models.

Morphine has little effect on wide dynamic range neurons studied in decerebrate cats (LeBars et al. 1976b), apparently because tonic descending inhibition is already powerful in this preparation. During reversible spinal cord block, morphine's action becomes pronounced. Although this observation does not contradict the notion that morphine has an important action in activating descending inhibitory pathways in animals with an intact neuraxis, it does indicate that in addition to opiate mechanisms in the brain stem, there are also opiate mechanisms at the spinal cord level that may affect nociceptive transmission.

There is a rich feltwork of neural processes and terminals in the uppermost layers of the dorsal horn that contain enkephalin-like immunoreactive substance (Sar et al. 1978; Uhl et al. 1979). In addition, when colchicine is used to block axoplasmic transport, enkephalin-like immunoreactivity is also present in neuronal cell bodies in the dorsal horn, especially in laminae I and II (Glazer and Basbaum 1981).

Opiate receptors have also been demonstrated in the dorsal horn (Atweh and Kuhar 1977a; LaMotte et al. 1976), and the concentration of these is reduced to half following dorsal rhizotomy (LaMotte et al. 1976). In a slice preparation of the trigeminal spinal nucleus caudalis, Jessell and Iversen (1977) demonstrated that opiates reduced the release of substance P from trigeminal afferent fibers, suggesting the possibility that opiate mechanisms may operate in part by presynaptic inhibition of transmitter release from fine afferent terminals, since substance P is contained in small dorsal root ganglion cells (Hökfelt et al. 1975, 1976) that presumably give rise to fine afferent fibers. These observations on dorsal horn opiate receptors led to the proposal that opioid substances may act presynaptically on primary afferent terminals in the dorsal horn. Further evidence is required to be sure that substance P is associated with nociceptive afferents. Other evidence that opiates decrease transmitter release has been reported (McDonald and Nelson 1978; Mudge et al. 1979).

In animals having spinal transections, systemically administered morphine selectively depresses the responses of wide dynamic range cells to Aδ and C fiber volleys, but not to Aαβ volleys (Fig. 42; LeBars et al. 1976a, 1980). The depression is naloxone-reversible. Interestingly, morphine microinjected into the NRM can enhance the responses of wide dynamic range cells to C fiber input (LeBars et al. 1980). When morphine is applied by microiontophoresis in the vicinity of dorsal horn neurons, or given systemically in animals with spinal cord transections, the

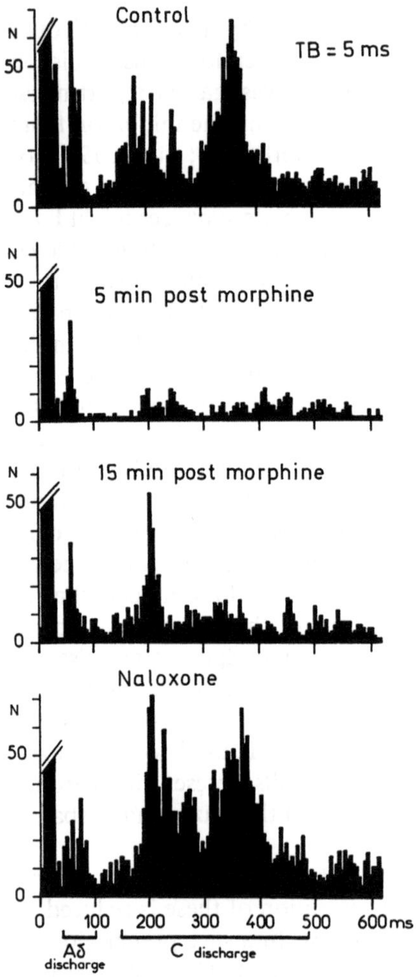

Fig. 42. Action of morphine on the responses of a dorsal horn interneuron to A and C fiber volleys. The *poststimulus time histograms* show the responses to Aαβ fibers (truncated), Aδ fibers, and C fibers. Systemic administration of morphine reduced the responses to the volleys in A and C fibers, but not those to Aαβ fibers. Since the spinal cord was transected, it can be presumed that the action of morphine was directly on the spinal cord. Naloxone (*bottom histogram*) reversed the morphine effect. Numbers indicate iontophoretic current. (LeBars et al. 1976a)

response of these cells to noxious stimuli, but not innocuous stimuli, may be blocked (Calvillo et al. 1974; LeBars et al. 1976a,b,c, 1979; Satoh et al. 1971); naloxone reverses this effect. A similar action has been shown for enkephalins (Fig. 43; Randić and Miletić 1978; Zieglgänsberger and Tulloch 1979). Opiates also affect the excitability of afferent terminals, including those of C fibers (Carstens et al. 1979b; Sastry 1978, 1979). In other experiments, opiates depress the responses to noxious and innocuous stimuli (Zieglgänsberger and Tulloch 1979). These actions may be reversed or prevented by naloxone (Fig. 43), and so they appear to depend upon an action on opiate receptors. At least part of the action of opiates appears to be on postsynaptic receptors (Zieglgänsberger and Bayerl 1976). A particularly interesting finding is that morphine or enkephalin may be effective when released in the substantia gelatinosa while recordings are made from a dorsal horn interneuron of the nucleus proprius (Duggan et al. 1976). One explanation of this observation might be that morphine is acting at presynaptic sites; another explanation is that the drug action may be on den-

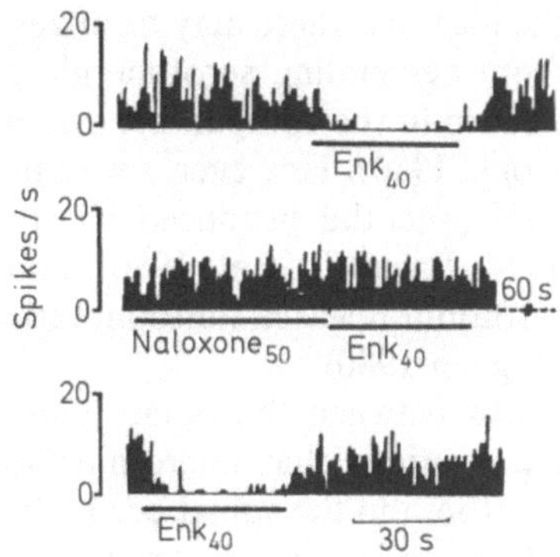

Fig. 43. Background discharge of a wide dynamic range neuron is depressed by microiontophoretic application of met-enkephalin (*upper and lower records*). Naloxone prevents this (*middle record*). Numbers indicate iontophoretic current. (Zieglgänsberger and Tulloch 1979)

drites ascending into the substantia gelatinosa from the neuron under investigation. More complex explanations are also conceivable.

Le Bars et al. (1981b) have found that systemic injections of morphine interfere with the DNIC mechanism. This is paradoxical, since naloxone was shown by the same investigators to interfere with DNIC (LeBars et al. 1981a) and since morphine is thought to act in part by increasing descending inhibition by activation of raphe- or reticulospinal neurons. Furthermore, this enhanced inhibition by morphine is supposed to be the most important component of opiate analgesia (see Section 2). However, LeBars et al. (1981b) suggest that the contrast between the activity of wide dynamic range neurons activated by a noxious input and the population of similar cells that are silenced by DNIC provides the message of the brain that signals pain. If this is true, then a reduction in the efficacy of DNIC by morphine would blur this message and result in analgesia.

Presuming that the raphe-spinal pathway acts at least in part on dorsal horn interneurons by the release of serotonin (Bourgoin et al. 1980; Bowker et al. 1981; Dahlström and Fuxe 1964), it is of interest that serotonin applied iontophoretically in the vicinity of dorsal horn interneurons or in the substantia gelatinosa tends to depress the activity of many cells, including their responses to noxious stimuli; however, some interneurons are excited (Belcher et al. 1978; Engberg and Ryall 1966; Griersmith and Duggan 1980; Headley et al. 1978; Randić and Yu 1976; Weight and Salmoiraghi 1966). Furthermore, inhibition following stimulation in the NRM is much less prominent after pretreatment with *p*-chlorophenylalanine, a serotonin synthesis blocker (Rivot et al. 1980). The serotonin receptor blocker LSD interferes with the inhibition of interneurons by stimulation in the PAG near the dorsal raphe nucleus (Guilbaud et al. 1973), as does methysergide (Carstens et al. 1981b). However, this effect could be due to antagonism of a serotonin synapse within the brain stem. Serotonin receptor blockers have been unable to block the inhibition produced by stimulation in the NRM in several studies, although the excitatory actions of such stimulation are blocked (Belcher et al. 1978; Griersmith et al. 1981). One possible explanation is that the receptors for serotonin in the CNS appear to be of at least two varieties, one commonly associated with inhibition and the other excitation. The commonly used serotonin receptor blockers are much less effective in interfering with the inhibitory than with the excitatory actions of serotonin (Haigler and Aghajanian 1974; Peroutka et al. 1981; Peroutka and Snyder 1979). Evidence consistent with the idea that there may be different kinds of serotonin synapses in the spinal cord from descending serotonergic pathways are the morphological differences described in the comparable synapses of the dorsal horn of the medulla (Ruda and Gobel 1980). However, serotonin receptor blockers do antagonize the late dorsal root potential produced by stimulation in the NRM (Proudfit and Anderson 1974; Proudfit et al. 1980), and they prevent the depression of interneurons by serotonin released iontophoretically in the substantia gelatinosa (Griersmith and Duggan 1980).

Links between the opiate and the serotonin mechanisms are provided by the observations that microinjection of morphine into the PAG causes release of serotonin in the spinal cord (Yaksh and Tyce 1979), and that systemic morphine causes an increase in serotonin synthesis in the spinal cord (Weil-Fugazza et al. 1981). Dickenson et al. (1981) present evidence that serotonin is involved in the DNIC of the activity of dorsal horn interneurons. Diffuse noxious inhibitory controls were reduced, although not eliminated, following pretreatment with *p*-chlorophenylalanine to deplete serotonin.

Catecholaminergic pathways may also influence the activity of dorsal horn interneurons. As mentioned earlier, there are catecholaminergic projections to the spinal cord from the brain stem (Blessing and Chalmers 1979; Carlsson et al. 1964; Dahlström and Fuxe 1964, 1965; Ross et al. 1981; Westlund and Coulter 1980; Westlund et al. 1981). Although some of the descending catecholaminergic axons may contain dopamine or epinephrine, the bulk of them are noradrenergic. The noradrenergic descending projections originate from the dorsolateral pons, including the nucleus locus ceruleus and subceruleus (Hancock and Fourgerousse 1976; Westlund and Coulter 1980; Westlund et al. 1981). Stimulation in the locus ceruleus causes inhibition of many dorsal horn interneurons (Hodge et al. 1981).

When norepinephrine is released adjacent to dorsal horn interneurons by microiontophoresis, the interneurons are generally depressed (Belcher et al. 1978; Biscoe et al. 1966; Engberg and Ryall 1966; Headley et al. 1978; Weight and Salmoiraghi 1966). The responses of interneurons to noxious stimuli are reduced in preference to the responses to innocuous stimuli (Headley et al. 1978). These observations are consistent with the pharmacological studies of the action of dopa on spinal cord reflex pathways (see Section 4), although they appear to conflict with other pharmacological evidence concerning the analgesia systems (Akil and Liebeskind 1975). However, the effect of systemic administration of drugs affecting catecholaminergic transmission could be complex, since there may be multiple sites of action of these agents giving apparently contradictory results. For example, Hammond et al. (1980) have found that it is possible to produce analgesia by local injections of phentolamine, an α-adrenergic blocking agent, into the NRM, suggesting a tonic inhibition of cells in the NRM by a catecholaminergic pathway. Systemic administration of an α-adrenergic blocker might therefore increase the amount of inhibition produced by the NRM of spinal cord neurons, thus enhancing analgesia.

A schematic drawing of the descending inhibitory pathways from the PAG, NRM, and other brain stem nuclei, along with possible neurotransmitter mechanisms, is shown in Fig. 44. This scheme was proposed by Basbaum and Fields (1978).

The role of other transmitters in the descending control systems affecting dorsal horn interneurons has been investigated to only a limited extent. The antagonists

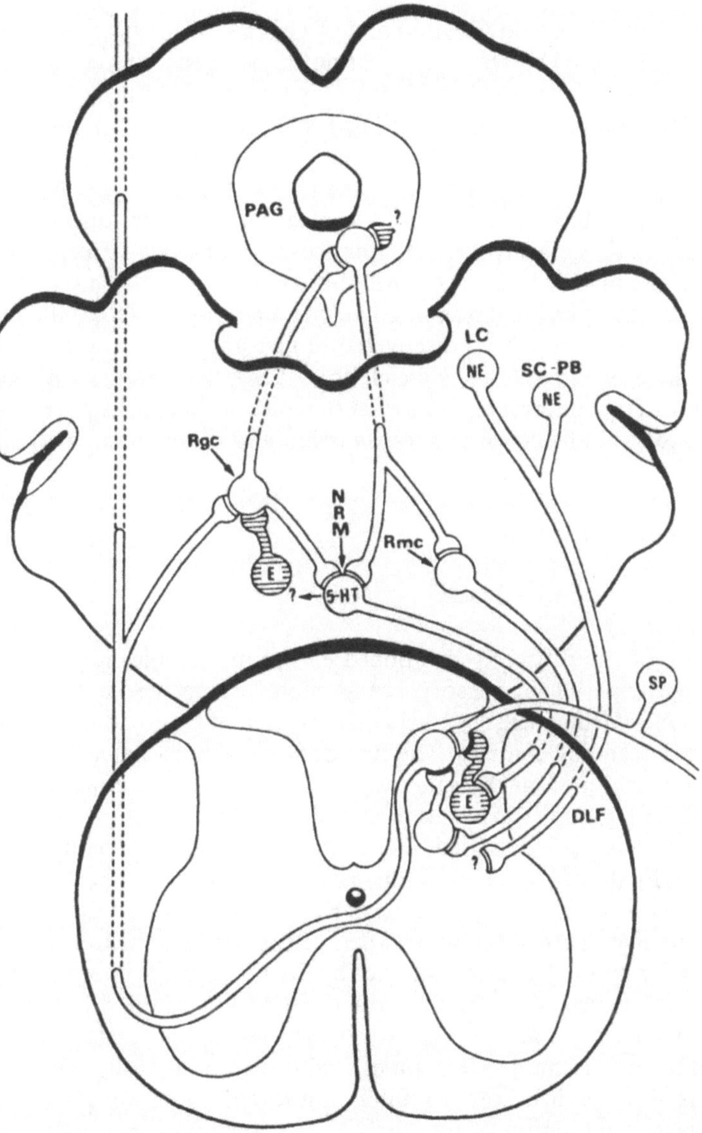

Fig. 44. Key elements in the endogenous system for control of nociceptive transmission. Neuron in the periaqueductal gray (*PAG*) (*upper section*) is shown receiving an input from a synaptic terminal that contains an endogenous opioid substance E (*hatched*). There is also an input from the reticular formation (R_{gc}, nucleus reticularis gigantocellularis). The PAG cell synapses on neurons in the rostral medulla (*middle section*) in the nucleus raphe magnus (*NRM*) and magnocellular reticular formation (R_{mc}). Cells in these nuclei provide descending serotonergic and nonserotonergic projections to the spinal cord (*lower section*) through the dorsal lateral funiculus (*DLF*). Noradrenergic projections (*NE*) to the cord from the nucleus locus ceruleus (*LC*) and subceruleus-parabrachial region (*SC-PB*) are also shown, as well as connections of the R_{gc} with NRM and interneurons containing opioid substances. Elements of the neural circuitry of the dorsal horn are indicated, including a substance-P-containing primary afferent (*SP*) and interneurons with or without endogenous opioids. Finally, an ascending tract cell is shown. (Basbaum and Fields 1978)

of γ-aminobutyric acid (GABA) and of glycine, bicuculline and strychnine, have been administered to see whether or not these agents affect the descending control systems. It appears that GABA may be involved but not glycine (Belcher et al. 1978; however, cf. Duggan et al. 1981).

Proudfit et al. (1980) have shown that the late dorsal root potential (DRP) produced by stimulation restricted to the raphe magnus region is reduced by serotonin receptor blockers, but not by semicarbazide. They conclude that serotonin, but not GABA, is involved in this form of primary afferent depolarization. Bicuculline also reduced the DRP, but they suggest this may be a nonspecific effect on the chloride ionophore likely to be responsible for the DRP.

Many peptides have been discovered in neurons that project to the spinal cord from the brain stem (e.g., Hökfelt et al. 1979; Swanson and McKellar 1979). It will be of great interest to learn what the roles of these substances are with respect to the control of nociceptive transmission and other functions.

5.2 Sensory Tract Cells

5.2.1 FRA-Activated Tracts

Because of the difficulty mentioned earlier in distinguishing between interneurons participating in sensory versus motor responses, it seems reasonable to determine whether sensory tracts that are excited by noxious stimuli are also subject to descending controls. The first demonstrations that sensory tract cells were tonically inhibited in the same way as dorsal horn interneurons were done by Lundberg's group (Carpenter et al. 1965; Holmqvist et al. 1960). The axons of the tract cells were identified by their location in the cord white matter and by excitation by the FRA. There is one pathway in the dorsal part of the lateral funiculus that is activated by ipsilateral FRA (Lundberg and Oscarsson 1961). Two other pathways are in the ventral part of the cord and include one path activated bilaterally by the FRA and another by the FRA of the contralateral side (Lundberg and Oscarsson 1962). The exact destination of the different tracts was not documented in detail. The FRA-activated ascending tracts were found to show an enhanced discharge after spinal cord transection (Holmqvist et al. 1960). The source of the tonic descending inhibition seems to be the medulla, since there was a release of inhibition of the tracts after a low pontine section (Carpenter et al. 1965).

5.2.2 Spinothalamic Tract

The spinothalamic tract (STT) is thought to be the main sensory tract in humans concerned with pain sensations originating from the body and viscera (Foerster and Gagel 1932; Kuru 1949; White and Sweet 1955, 1969). The STT is also likely to have the same role in subhuman primates (Vierck and Luck 1979; Yoss 1953), but it seems to be less crucial for nociception in the cat (Kennard 1954). The responses can be recorded from neurons identified as STT cells by antidromic activation from the thalamus (Albe-Fessard et al. 1974; Applebaum et al. 1975;

Dilly et al. 1968; Giesler et al. 1976; Trevino et al. 1972, 1973). Thus it is possible to study the responses of cells identified as belonging to a candidate nociceptive sensory tract and to determine whether these responses are appropriate for nociception and if these cells are under descending control.

Spinothalamic tract cells can be classified in the same fashion as dorsal horn interneurons in general (Chung et al. 1979; McCreery et al. 1979a,b; Price et al. 1978; Willis et al. 1974). The nociceptive spinothalamic cells include both high-threshold and wide dynamic range cells (Fig. 45; Chung et al. 1979; Willis 1981, to be published). The responses of spinothalamic cells to graded intensities of noxious mechanical or thermal stimuli applied to the skin are appropriate for a role of these neurons in nociception (Chung et al. 1979; Kenshalo et al. 1979, 1982; Willis et al. 1974, 1975). Furthermore, many of them have a convergent input from skin, muscle, and/or viscera and so may be involved in the phenomenon of pain referral (Foreman et al. 1977, 1979a,b, 1981; Foreman and Weber, 1980; Hancock et al. 1975; Levante et al. 1975; Milne et al. 1981). Many of the properties of spinothalamic cells have been reviewed elsewhere (Willis and Coggeshall 1978; Willis to be published).

There has not been a thorough study of the effects of spinal cord transection on the responses of STT cells to noxious stimuli. Willis et al. (1977) showed that there was an enhancement of the responses of a few STT cells after lesion of the dorsal part of the lateral funiculi, and Gerhart et al. (1981b) found that the background discharges of spinothalamic cells were often enhanced after spinal cord

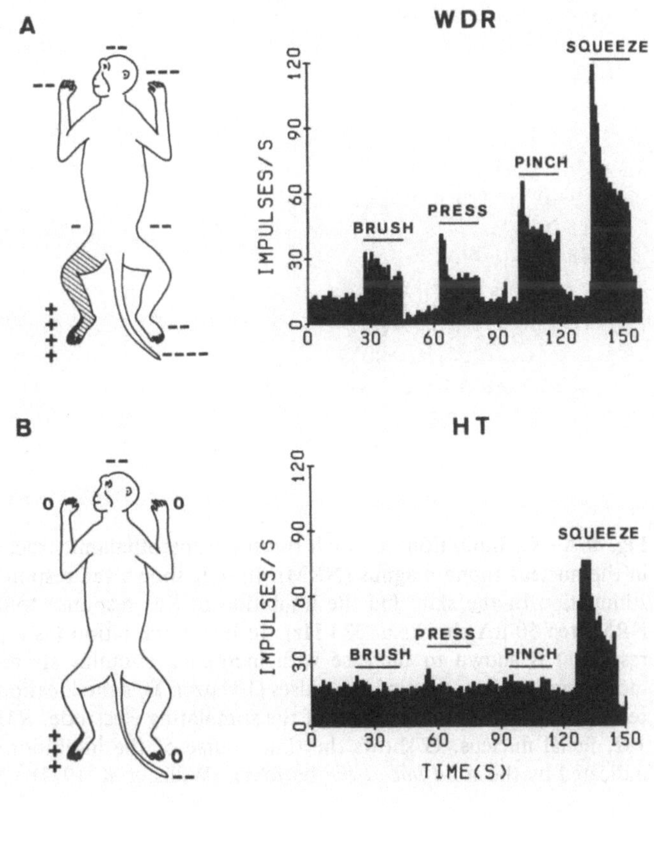

Fig. 45A, B. Receptive fields and responses of spinothalamic tract (STT) neurons. In **A** are shown *on the left* the excitatory and inhibitory receptive fields of a wide dynamic range (*WDR*) STT cell. The *histogram* shows the responses of this cell to innocuous brushing of the excitatory field, pressure, pinch, and squeezing. In **B**, similar results are shown for a high-threshold (*HT*) STT cell. (Willis 1981)

transections. These observations suggest that there is a tonic descending control system at work on STT cells. Whether this tonic inhibition is enhanced in the decerebrate state is not known, since STT cells cannot be identified after decerebration.

Stimulation in the PAG inhibits not only interneurons in general, but also STT cells in particular (Hayes et al. 1979c; Yezierski et al. 1982). Both high-threshold and wide dynamic range cells are inhibited (see Fig. 51; Yezierski et al. 1982).

In addition to the PAG, STT cells can be inhibited by stimulation in the ventral posterior lateral nucleus of the thalamus (Gerhart et al. 1981c) and the sensorimotor cortex (Coulter et al. 1974; Yezierski et al. 1981). Using brief stimulus trains, cortical stimulation causes a selective inhibition of tactile responses of spinothalamic cells, leaving nociceptive responses intact (Coulter et al. 1974). However, with long stimulus trains, cortical stimulation can inhibit nociceptive, as well as nonnociceptive, responses (Yezierski et al. 1981). Stimulation of some areas of cortex excites spinothalamic cells (Coulter et al. 1974; Yezierski et al. 1981).

Because of the finding that stimulation of the NRM causes an inhibition of spinal cord interneurons, it is of interest that comparable stimulation has an inhibitory

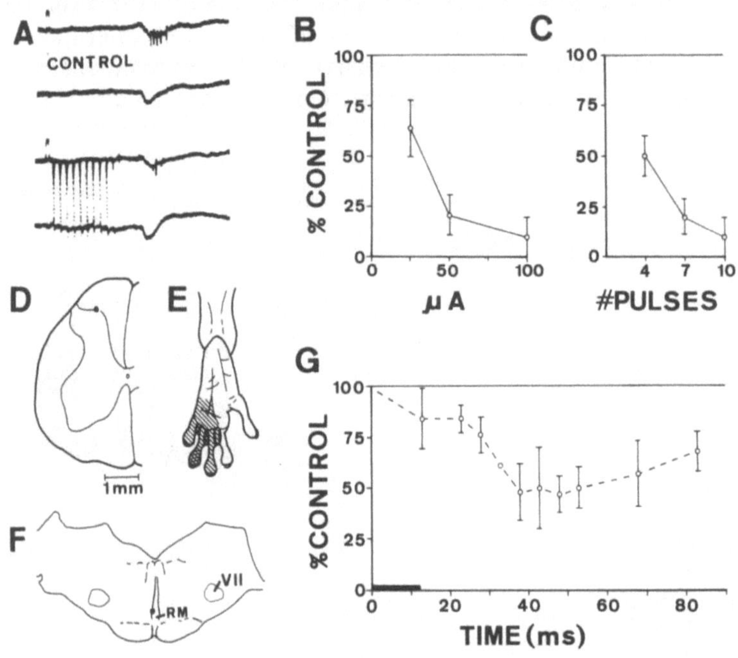

Fig. 46A – G. Inhibition of a high-threshold spinothalamic tract (STT) cell by stimulation in the nucleus raphe magnus (NRM). In **A** is shown the response of the cell to electrical stimulation of the skin and the inhibition of this response following stimulation in the NRM (ten 50-μA shocks at 333 Hz). In **B**, the inhibition (as a percentage of the control response) is shown to increase with increasing stimulus strength (ten stimuli) or with increasing numbers of stimulus pulses (100 μA). **D** is the location of the STT cell; **E** is the receptive field; **F** is the location of the stimulating electrode. *RM*, nucleus raphe magnus; *VII*, facial nucleus. **G** shows the time course of the inhibition (stimulus train duration indicated by the *solid bar at the bottom*). (Willis et al. 1977)

effect on STT cells (Fig. 46; Beall et al. 1976; Gerhart et al. 1981a; McCreery and Bloedel 1975; McCreery et al. 1979b; Willis et al. 1977; Yezierski et al. 1982). Some features of the inhibitory effects of NRM stimulation on STT cells include the following. The inhibition improves with repetitive stimulation, suggesting that temporal summation is needed (in anesthetized preparations) for a full manifestation of the inhibition (Willis et al. 1977). The threshold for the inhibition can be less than 25 μA, suggesting that the stimulus need not spread outside the bounds of the raphe to have an effect (Willis et al. 1977). However, only electrical stimuli have so far been used, and so it is not certain that the effects are due to the excitation of neurons within the NRM or to axons of passage. The inhibition increases as the stimulus strength is raised (Willis et al. 1977). With high-stimulus currents, it is not clear to what extent elements outside the raphe proper are involved in the inhibition. Excitatory actions have also been reported (McCreery et al. 1979b; Yezierski et al. 1982). These were common in the study by McCreery et al. (1979b), who used single shocks of high intensity, but rare in the experiments of the Willis group (1977), who used repetitive stimulation at lower intensities.

The inhibitory pathway from the NRM to STT cells is not known with certainty. The latency of the initial inhibition is about 15 ms, suggesting that a myelinated conducting system (at least 10 m/s) is required to explain the onset of the inhibition (Willis et al. 1977). More slowly conducting components of the pathway could, of course, contribute to later parts of the inhibition. The inhibition produced by brief trains of NRM stimuli can be abolished by lesions of the dorsal parts of the lateral funiculi, suggesting that the raphe-spinal tract may be involved (Basbaum et al. 1978; Basbaum and Fields 1979; Martin R.F. et al. 1978; Tohyama et al. 1979a,b; Watkins et al. 1980; Willis et al. 1977). However, a role of reticulospinal projections from the magnocellular part of the nucleus gigantocellularis cannot be ruled out, especially in cases in which high-stimulus currents are used or the stimulating electrode is placed to one side of the midline (Basbaum et al. 1978; Tohyama et al. 1979a,b). Furthermore, indirect projection from other parts of the brain stem, after ascent of volleys from the NRM, cannot be ruled out, especially in the case of later components of the inhibition.

The inhibition has been tested against the responses of STT cells to various types of input. With respect to inputs evoked by electrical stimulation of cutaneous nerves, the inhibition appears to be preferentially directed at the responses to Aδ and C fibers and less powerful on the responses to Aαβ fibers (Fig. 47; Gerhart et al. 1981a; Willis et al. 1977; see Beall et al. 1977; Foreman et al. 1975). Although this would lead to the expectation that there should be a more powerful inhibition of the responses of STT cells to noxious than to innocuous stimuli, this is only rarely the case. Usually, the responses of STT cells to brushing the hair or skin are as powerfully inhibited as are the responses to noxious pinch or heat (Fig. 48E – H; Gerhart et al. 1981a; Willis et al. 1977). However, in some cases there is a differential inhibition of responses to noxious stimuli (Fig. 48A – D; Gerhart et al. 1981a). It should be kept in mind that the Aδ fibers in the monkey supply not only nociceptors but also D hair follicles (Perl 1968). In the study by McCreery and Bloedel (1975), the responses of STT cells to slowly rising, strong mechanical stimuli were more affected than were the transient responses at the onset of stimulation.

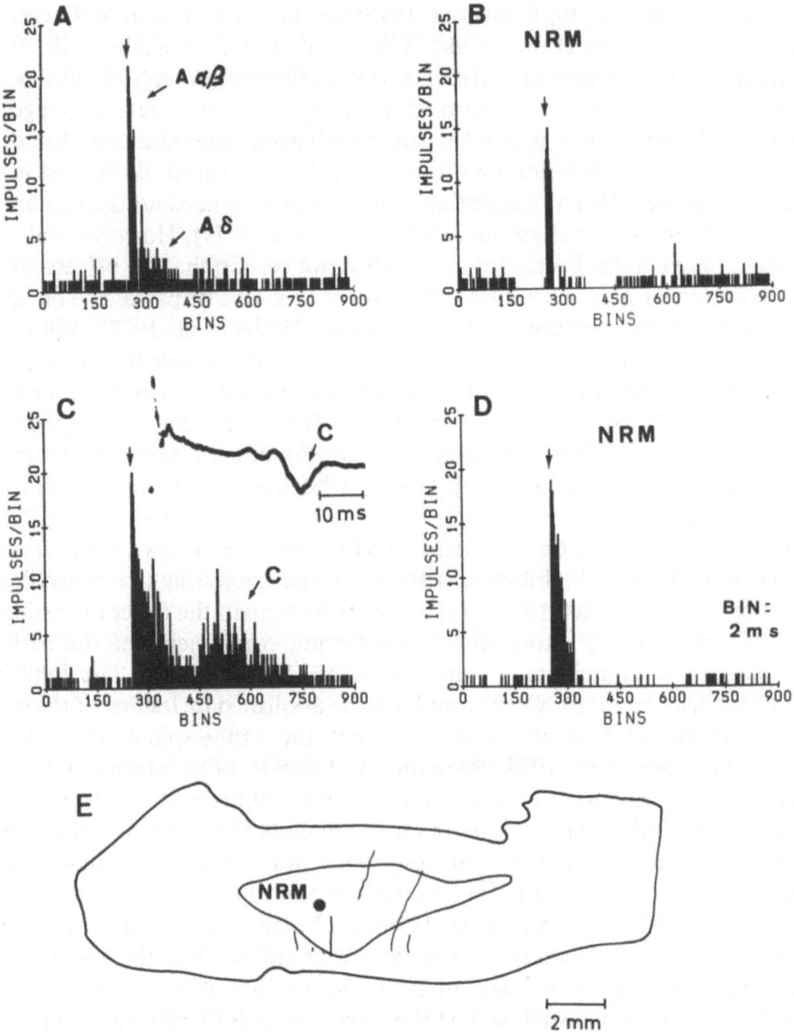

Fig. 47A – E. Inhibition of the responses of a wide dynamic range spinothalamic tract cell to stimulation of A or C fibers of the sural nerve by descending volleys initiated by stimulation in the nucleus raphe magnus (*NRM*). The histograms in **A** and **C** show the control responses to stimulation of A and C fibers, respectively. The *inset* in **C** is the C fiber volley recorded from the sural nerve. Stimulation in NRM during the times shown by the bars in **B** and **D** (150-μA pulses at 333 Hz) inhibited the responses. **E** shows the location of the stimulating electrode in a sagittal section at the midline of the brain stem. (Gerhart et al. 1981a)

The mechanism of inhibition by NRM stimulation appears to be complicated. Nucleus raphe magnus stimulation causes primary afferent depolarization of myelinated afferent fibers (Martin R.F. et al. 1979; Proudfit and Anderson 1974; Proudfit et al. 1980). The fibers that are affected include large and small myelinated fibers that innervate sensitive mechanoreceptors, and small myelinat-

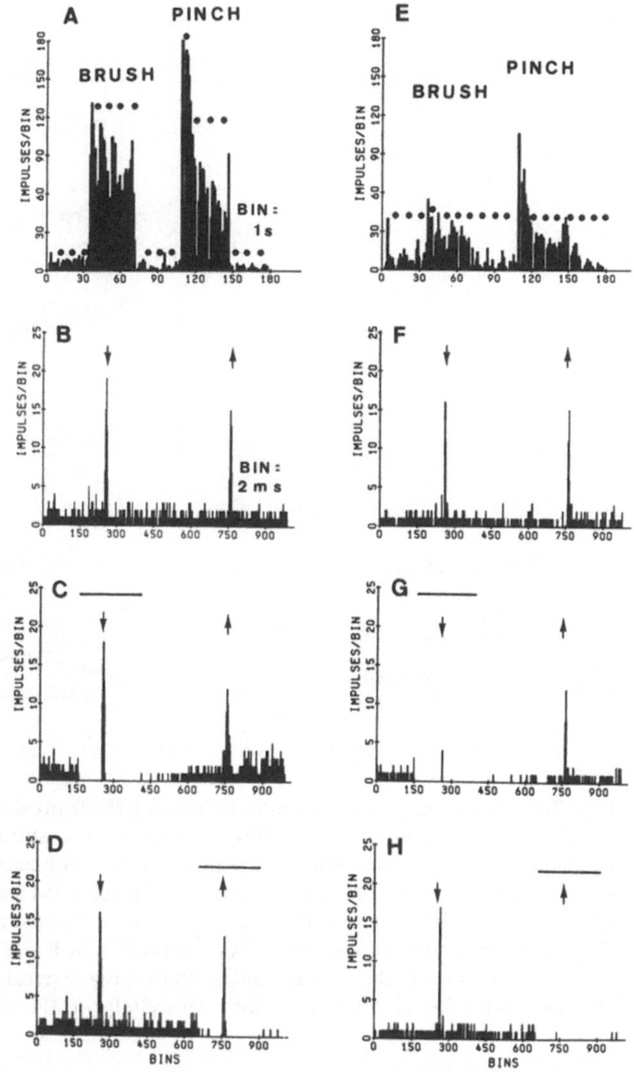

Fig. 48A – H. Preferential inhibition of nociceptive responses of an STT cell and lack of preferential inhibition of another following stimulation of the nucleus raphe magnus (NRM). The cell in **A – D** demonstrated a selective inhibition, whereas that in **E – H** did not. The histograms in **A** and **E** show responses of these wide dynamic range cells to brushing and to pinching. At the times indicated by the *dots*, the NRM was stimulated (2-s trains, 200 μA, 333 Hz). **B** and **F** show the control responses to indentation of the skin with a servo-controlled stimulator (1-s indentations of 0.5 mm from initial static indentation of 0.5 mm, repeated ten times at 5-s intervals). In the other records, the NRM was stimulated at the times shown by the *bars* (150-μA pulses, 333 Hz). (Gerhart et al. 1981a)

ed fibers that innervate nociceptors (Martin R.F. et al. 1979). Nucleus raphe magnus stimulation also affects the excitability of unmyelinated afferent fiber terminals (Hentall and Fields 1979). However, in this case the fibers show a reduced excitability, suggesting primary afferent hyperpolarization. It is not clear whether primary afferent hyperpolarization in such a case would necessarily lead to presynaptic facilitation. In addition to a possible presynaptic site of action, the effects of NRM stimulation on STT cells are also mediated in part by a postsynaptic inhibition (Fig. 49; Giesler et al. 1981a; cf. Engberg et al. 1968d). The finding that STT cells can generate inhibitory postsynaptic potentials in response to NRM stimulation helps explain why there should often be a nonselective inhibition of the response both to innocuous and to noxious stimuli. Perhaps those cells that show a preferential inhibition of noxious inputs are ones that are

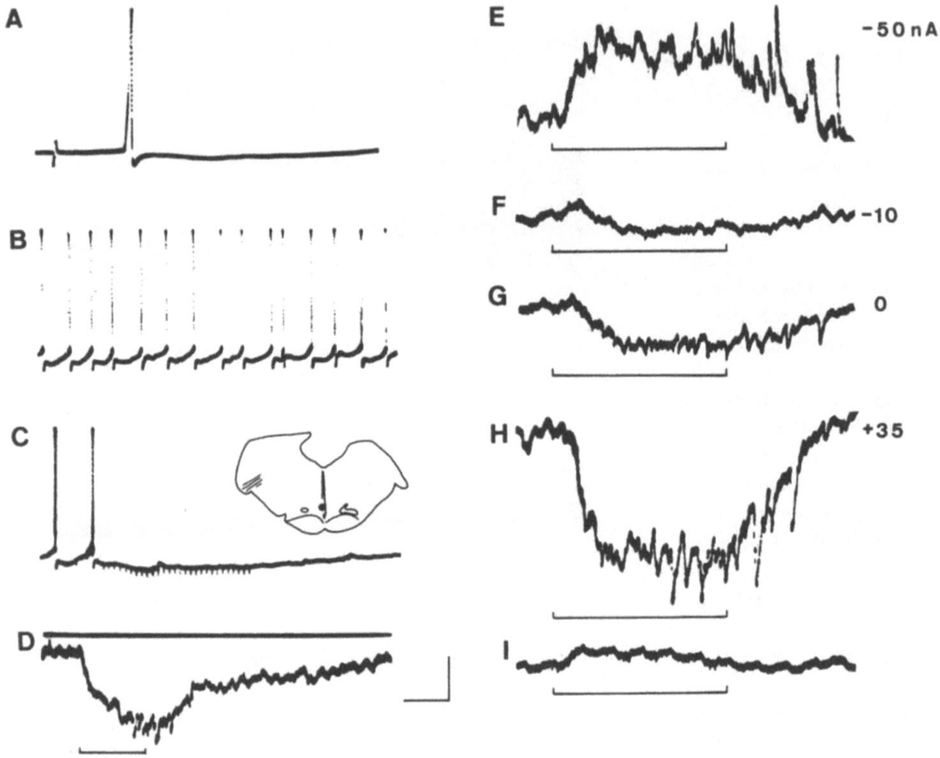

Fig. 49A – I. Inhibitory postsynaptic potential (IPSP) produced in a spinothalamic tract (STT) cell by stimulation in the NRM. **A** shows the antidromic spike resulting from stimulation in the contralateral ventral posterior lateral nucleus of the thalamus. **B** shows background discharges that in **C** are inhibited during NRM stimulation (100-ms train, 200-μA pulses, 333 Hz). The stimulus site is indicated in the *inset*. The high-gain record in **D** shows the time course of the IPSP. For the records in **E – H**, the membrane potential of the cell was altered by the currents indicated to show reversal of the IPSP. The record in **I** was taken with the microelectrode just extracellular to the cell. (Giesler et al. 1981a)

subject only to presynaptic inhibition and not postsynaptic inhibition. However, as mentioned above, NRM stimulation causes primary afferent depolarization of mechanoreceptor afferents, as well as of nociceptor afferents.

The effects of NRM stimulation appear to be mediated in part, at least, by serotonin. Evidence that indirectly supports this is the finding that iontophoretic application of serotonin in the vicinity of STT cells causes a depression of their background activity and also of their responses to noxious stimuli and to glutamate ions (Jordan et al. 1978, 1979). Interestingly, deep STT cells are often excited by iontophoretically released serotonin (Jordan et al. 1979), and so the depression of high-threshold and wide dynamic range STT cells is probably not just a nonspecific action of serotonin. It may be significant that serotonin depresses the responses of STT cells to iontophoretic pulses of glutamate (Fig. 50; Jordan et al. 1978). If the glutamate excites STT cells strictly through a postsynaptic action on the membranes of the STT cells, then it would appear that

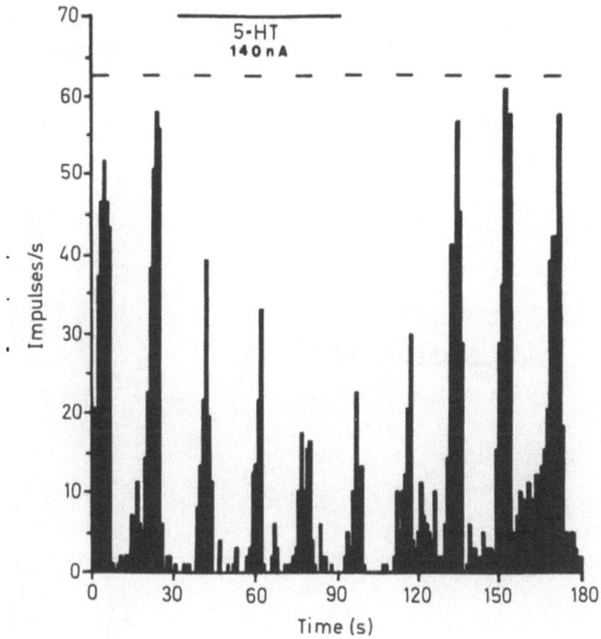

Fig. 50. Depression of the responses of a spinothalamic tract cell to glutamate pulses by serotonin (*5-HT*, 5-hydroxytryptamine). The cell was a wide dynamic range neuron, and it was excited by a series of pulses of glutamate released by micro-iontophoresis from a multi-barreled micropipette (times indicated by the *short bars*). When serotonin was released as well (*long bar*), the responses to glutamate were diminished. (Jordan et al. 1978)

serotonin has at least in part a postsynaptic action. Such an action would be consistent with the finding of serotonin-containing synaptic endings on many neurons in the superficial layers of the dorsal horn, layers that contain many cells projecting to the thalamus (Ruda and Gobel 1980). However, it has yet to be demonstrated that serotonin terminals are located on identified STT cells.

Further evidence for a serotonergic action of the descending volleys caused by NRM stimulation comes from experiments in which serotonin receptor antagonists were used to interfere with the NRM action (Yezierski et al. 1982). When the NRM was stimulated with 200-ms trains of pulses at strengths of 200 μA, serotonin blockers produced a significant reduction in the inhibition. However, when 2-s stimulus trains were used at a similar stimulus intensity, the serotonin blockers did not have a significant effect (Fig. 51). Several possible explanations were suggested (Yezierski et al. 1982). The long stimulus trains might activate other descending pathways than a serotonergic raphe-spinal pathway; since serotonin receptor blockers may not be completely effective against CNS inhibitory serotonergic synapses (Haigler and Aghajanian 1974; Peroutka et al. 1981; Peroutka and Synder 1979), perhaps the longer trains release enough serotonin to prevent effective blockade; or perhaps the raphe-spinal neurons release additional transmitters, such as substance P, when long trains are employed (cf. Chan-Palay 1979; Hökfelt et al. 1978).

It is interesting that serotonin receptor blockers produce a more complete disruption of PAG inhibition than of NRM inhibition (Yezierski et al 1982). One explanation may be that there is another, perhaps excitatory, serotonergic synapse between the PAG and a bulbar relay, for example in the NRM or reticular formation (cf. Bradley and Wolstencroft 1965; Briggs 1977; Couch 1970). Another possibility is that stimulation in the PAG causes the excitation of

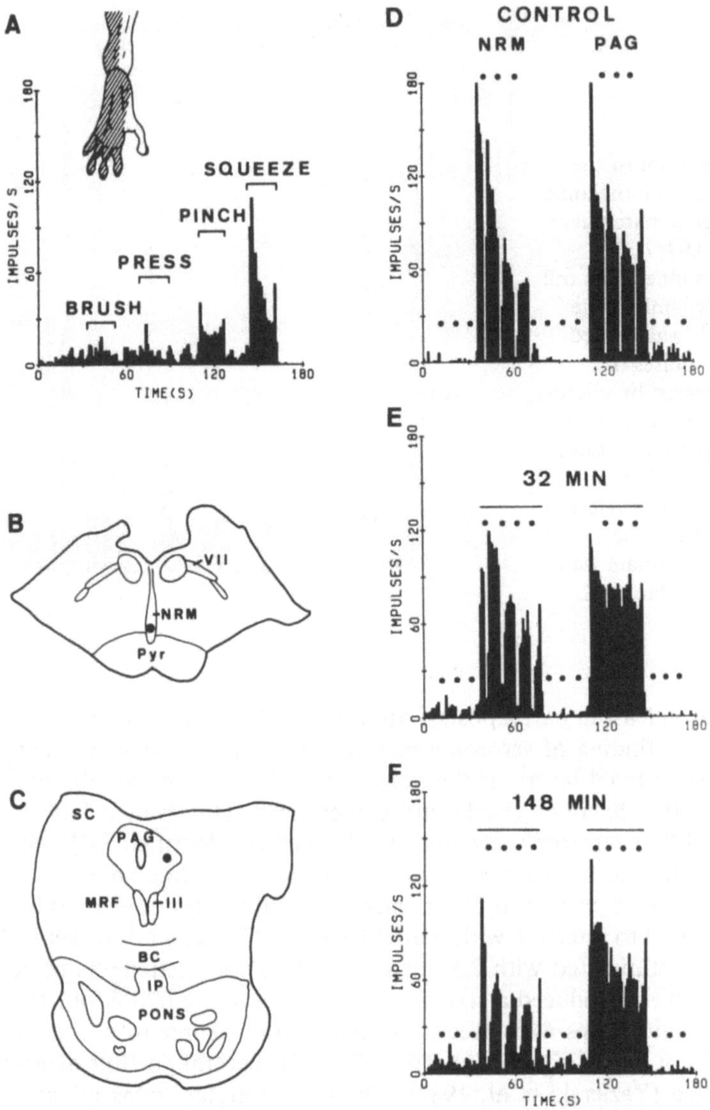

Fig. 51A–F. Inhibitory effect of stimulation in the periaqueductal gray (*PAG*) and nucleus raphe magnus (*NRM*) on the responses of a wide dynamic range *spinothalamic tract* cell to noxious mechanical stimuli, and the action of a serotonin receptor blocker, methysergide, on the inhibition. The responses of the cell to graded mechanical stimuli are shown in **A**, along with the receptive field location. **B** and **C** show the stimulation sites in the NRM and PAG. *BC*, brachium conjunctivum; *MRF*, midbrain reticular formation; *SC*, superior colliculus; *III*, oculomotor nucleus. In **D**, stimulation of either the NRM or the PAG at the times indicated by the *dots* results in a powerful inhibition of the responses to pinching the skin. In **E**, intravenous administration of methysergide (5 mg/kg) results in a substantial block of the inhibitory action of PAG stimulation without producing much of a change in the NRM inhibition. **F** shows partial recovery. (Yezierski et al. 1982)

a purely serotonergic component of the NRM, in contrast to stimulation in the NRM, which may activate additional nonserotonergic pathways (Yezierski et al. 1982).

Spinothalamic tract cells can also be inhibited by stimulation in other regions of the brain stem. For example, many STT cells are inhibited by stimulation in the reticular formation, including the region of the nucleus gigantocellularis (Gerhart et al. 1981a; Haber et al. 1978, 1980). The reticular inhibition is very similar to that produced by NRM stimulation, both in time course and in being nonselectively effective against the responses to both innocuous and noxious stimulation (Gerhart et al. 1981a; Haber et al. 1980).

Descending excitation has also been demonstrated in the case of reticular formation stimulation (Haber et al. 1980). Excitation was also commonly observed preceding inhibition from stimulation either of the reticular formation or in the NRM by McCreery et al. (1979b).

The STT cells so far described have been identified by antidromic activation from the lateral thalamus, in or near the ventral posterior lateral nucleus (Trevino et al. 1973). They therefore belong to one part of the STT; another component of the STT projects to the medial thalamus, with terminals especially concentrated in the central lateral nucleus (Berkley 1980; Boivie 1979; Bowsher 1961; Kerr 1975; Mehler et al. 1960; Willis et al. 1979). Recently, a study has been done on the response properties of STT cells that can be backfired from the medial thalamus, and these properties were compared with those of the laterally projecting STT cells (Giesler et al. 1981b). Actually, three groups of STT cells could be identified: STT cells projecting just to the lateral thalamus; those projecting to lateral and medial thalamus; and those projecting just to the medial thalamus. The first two groups had similar response properties. The third group that projected just to the medial thalamus was quite different. These cells were often high-threshold cells, and they commonly had whole-body-excitatory receptive fields. The cells were all sampled in the lumbosacral enlargement, yet they could be excited, for example, by noxious stimulation of the forelimbs or face, as well as the hind limbs and tail. The excitation from the ipsilateral hind limb was preserved after transection of the upper cervical spinal cord, but that from the remainder of the body and the face was eliminated, suggesting that much of the excitatory receptive field depended upon a pathway that looped through the brain. Stimulation in the reticular formation caused a potent excitation of these cells, although some locations also had an inhibitory action. No systematic investigation of the descending control systems operating on these cells has been done, but it is suggested that the interconnections with the reticular formation are likely to be very important (Giesler et al. 1981b), and part of the inhibition is likely to be mediated by descending control paths.

Inhibition of STT cells from stimulation of peripheral structures has also been studied. McCreery and Bloedel (1976, 1979a) found that stimulation of mechanoreceptors in the skin of the face or of large myelinated afferents in the trigeminal nerve of the cat resulted in inhibition of STT cells. These effects of trigeminal afferents are undoubtedly relayed through the brain stem. Details of the descending pathway were not examined.

Occasionally, STT cells can be inhibited by tactile stimuli, such as hair movement (e.g., Milne et al. 1981), although usually this is not the case (Gerhart et al.

1981b). Inhibition of STT cells can also be produced by stimulation of the dorsal column of the spinal cord (Foreman et al. 1976). The inhibition is due to segmental actions by the antidromically conducted volley in the dorsal column, since inhibition is not produced in STT cells by stimulation just rostral to a lesion interrupting the dorsal column, whereas inhibition is produced by stimulation just caudal to such a lesion.

Intense stimulation of the skin over much of the body and face or of small myelinated and unmyelinated afferents of a peripheral cutaneous nerve will inhibit many STT cells projecting to the lateral thalamus, both of the wide dynamic range and of the high-threshold classes (Fig. 45; Gerhart et al. 1981b). This effect resembles that of the DNIC system described by Besson's group (LeBars et al. 1979a,b). One difference is that DNIC has little effect on high-threshold dorsal horn interneurons (LeBars et al. 1979b; however, cf. Gray and Dostrovsky 1981). A further difference is that a substantial part of the inhibitory action, especially from the contralateral hind limb and the tail, is preserved after transection of the spinal cord at an upper cervical level (Gerhart et al. 1981b). By contrast, DNIC is eliminated by spinal cord transection (LeBars et al. 1979a). The inhibition of STT cells from noxious stimulation of the forelimbs is reduced by spinal transection, and so part of the inhibition may depend upon a supraspinal loop, as in the case of DNIC. The fact that many neurons in the NRM and in the reticular formation that project to the spinal cord are excited by noxious stimulation of widespread areas of the body surface is consistent with the hypothesis that part of the wide noxious inhibitory receptive fields of STT cells may result from excitation of descending control systems by noxious stimuli.

5.2.3 Spinocervical Tract

The spinocervical tract is largely concerned with signalling tactile information; however, cells in this pathway have also been reported to respond to nociceptive stimuli (Brown and Franz 1969; Brown 1971; Cervero et al. 1977a; Hong et al. 1979; Kniffki et al. 1977). Therefore, it is reasonable to consider the spinocervical tract as a candidate nociceptive pathway. There is no evidence that this pathway is important for nociception in humans or in subhuman primates, but Kennard's study (1954) is consistent with a role for this tract in nociception in the cat. Furthermore, since pain sometimes returns after initially effective cordotomies in humans (White and Sweet 1955), there are presumably alternative nociceptive pathways that become more important after interruption of the STT. The spinocervical tract should be considered as one alternative for such a role in humans, since this pathway does seem to be present in the human spinal cord (Truex et al. 1965).

The spinocervical tract is clearly under the control of a tonic descending inhibition, since transection of the spinal cord increases the responses of spinocervical tract cells to noxious stimuli (Cervero et al. 1977a; Hong et al. 1979) and changes the response properties of certain cell classes in a direction indicating a more effective input from nociceptors (Brown 1971).

Spinocervical tract cells can be inhibited by stimulation in the brain stem (Taub 1964), and microstimulation in the white matter of the spinal cord reveals that certain zones contain axons that can inhibit spinocervical tract cells; the inhibitory fibers lie in the two dorsolateral funiculi and in the two ventral funiculi (Brown et al. 1973). However, it is not yet clear what the source of these fibers is. Stimulation of the sensorimotor cortex causes an inhibition of spinocervical tract cells (Brown et al. 1977).

Although there is some evidence that there are nociceptive cells in the lateral cervical nucleus (summarized in Willis and Coggeshall 1978), more studies of the response properties of these cells are needed. Of particular interest would be evidence of responses to noxious heat stimuli (cf. Cervero et al. 1977a) or to chemical stimuli (Hong et al. 1979; Kniffki et al. 1977).

5.2.4 Other Tracts

There are several other ascending sensory tracts that are candidates to play some role in nociception. One of these is the spinoreticular tract (Fields et al. 1975, 1977b), and another is the second-order dorsal column pathway (Angaut-Petit 1975; Uddenberg 1968). Neither of these pathways has been studied in detail from the perspective of the descending ceontrol systems.

5.3 Conclusions

1. Nociceptive signals are relayed to the brain by way of neurons in the dorsal horn. For this reason, a useful means of investigating the descending control of nociceptive transmission is by monitoring changes in the activity of dorsal horn neurons in response to stimulation of descending pathways.
2. Nociceptive dorsal horn interneurons include nociceptive-specific (high-threshold) cells and wide dynamic range cells. Both cell types are activated by input from nociceptors, but wide dynamic range cells are also activated by input from mechanoreceptors. Other nonnociceptive dorsal horn cells include low-threshold (excited by sensitive mechanoreceptors) and deep (excited by subcutaneous receptors) cells.
3. Classifications of dorsal horn cells according to their inputs from broad categories of mechanoreceptors and nociceptors are likely to give an overly simplistic view of the role of these cells. Such classification schemes have practical value, but inputs from particular kinds of sense organs tend to be overlooked. Another problem is that one cannot assume that a dorsal horn neuron classified as nociceptive-specific, for example, contributes to the sensation of pain. It may instead be a part of a neural circuit for a reflex response to noxious input.
4. Dorsal horn interneurons receiving nociceptive inputs are subject to a tonic inhibitory control, as shown by an enhanced responsiveness after spinal cord transection. This inhibitory action may originate in the medial part of the lower brain stem, since powerful inhibitory actions, involving postsynaptic inhibition in at least some cases, can be demonstrated in recordings from interneurons as a result of stimulation in the nucleus raphe magnus (NRM) and the adjacent reticular formation.

5. Stimulation in the periaqueductal gray (PAG) results in the inhibition of dorsal horn interneurons. There is a preferential action on responses to noxious as opposed to innocuous stimulation of the skin. Inhibition also results when the adjacent midbrain reticular formation is stimulated, but in this case there is a change in the threshold of the response, whereas PAG stimulation causes a change in the gain of the stimulus-response curve.
6. The nociceptive responses of dorsal horn interneurons are also inhibited following stimulation in the NRM. Whether or not this inhibition is selective is controversial. Inhibition can also be produced by stimulation in the adjacent reticular formation. The descending inhibitory pathway from the NRM is in the dorsal part of the lateral funiculi, and the fastest axons in the pathway are myelinated. The terminals of the axons of the raphe-spinal neurons are in the uppermost layers, as well as in the neck and base, of the dorsal horn.
7. Since there are few direct projections from the PAG to the spinal cord, it is believed that the inhibitory actions of PAG stimulation depend upon a relay from PAG in the lower brain stem. The most likely site of the relay is in the NRM and adjacent reticular formation. There are known PAG projections to these areas. Furthermore, PAG stimulation excites many NRM neurons, including raphe-spinal cells.
8. Since many neurons in PAG and NRM respond to noxious inputs, it has been suggested that the descending inhibitory system acts as a negative feedback loop that serves to limit nociceptive transmission.
9. Another concept for the function of the descending inhibitory pathways comes from the description of a widespread inhibitory receptive field of wide dynamic range neurons. The inhibition depends upon noxious stimulation anywhere on much of the body surface, and so the mechanism has been termed "diffuse noxious inhibitory controls" (DNIC). It has been proposed that DNIC serve to enhance contrast in the responses of the population of wide dynamic range neurons. Since the inhibitory action depends upon an ascending pathway to the brain stem that in turn activates the descending inhibitory pathways, the proposal is made that these inhibitory pathways contribute to the sensoridiscriminative processing of nociceptive information. Several features of the DNIC mechanism have been questioned, however.
10. Injections of morphine into the PAG or NRM may result in the inhibition of dorsal horn interneurons. Similar injections into PAG can activate (or inhibit) neurons of NRM. Naloxone has been reported to decrease the inhibition produced by stimulation of the NRM, although other investigators do not find that PAG inhibition is naloxone-reversible. These observations suggest that an opiate mechanism may contribute to PAG and NRM inhibition of dorsal horn interneurons, but not all the results are consistent.
11. Morphine may also have an action at the level of the spinal cord. There are enkephalin-containing terminals and also cell bodies in the dorsal horn, and there are opiate receptors, as well, some of which may be associated with primary afferent fibers. The latter may prevent the release of transmitter from afferent fibers. When systemic morphine is given to a cat with a spinal cord transection, the responses of wide dynamic range interneurons to inputs from Aδ and C fibers, but not to Aαβ fibers, are reduced. Morphine released

near interneurons by microiontophoresis causes a depression of the responses to noxious stimuli. At least part of the action of morphine appears to be postsynaptic.

12. Paradoxically, small doses of morphine reduce DNIC. It is suggested that this results in analgesia because of a reduction in the signal-to-noise ratio in the ascending nociceptive transmission pathways.
13. Inhibition by NRM may involve the release of serotonin. When serotonin is applied in the vicinity of dorsal horn interneurons by microiontophoresis, many of the cells are depressed, although some are excited. Nucleus raphe magnus inhibition is reduced after depletion of CNS serotonin by *p*-chlorophenylalanine. It has been difficult to demonstrate a reduction in NRM inhibition by administration of serotonin receptor blockers (although excitatory actions are blocked). This may be due to the nature of the serotonin receptor responsible for the inhibition.
14. Periaqueductal gray inhibition can be blocked by serotonin receptor blockers. One explanation of this would be that the descending pathway from the PAG involves an excitatory serotonergic synapse within the brain stem.
15. Catecholaminergic pathways may also influence the activity of dorsal horn interneurons. Norepinephrine released microiontophoretically near dorsal horn interneurons depresses their activity, especially their responses to noxious stimuli. Systemic injections of agents that affect catecholaminergic synapses may produce complex effects if they act both at the level of the brain stem and spinal cord. Thus the evidence on the involvement of these pathways in the regulation of interneuronal activity is as yet unclear.
16. Other transmitter and modulator substances, including amino acid transmitters and peptides, may prove to have a role in the regulation of nociceptive responses of dorsal horn interneurons.
17. In addition to unidentified interneurons, the descending inhibition of ascending tract cells has been investigated. There are several ascending tracts that are activated by the flexion reflex afferents which have been shown to be under tonic descending inhibitory control in decerebrate animals. Their responses are enhanced after spinal cord transection.
18. The spinothalamic tract (STT) is a major ascending pathway that is likely to play a major role in pain sensation, at least in humans and in subhuman primates. The responses or STT cells can be classified in the same fashion as those of dorsal horn interneurons in general. Furthermore, many of the same features of descending inhibitory controls apply as well to STT cells as to interneurons. For example, STT cells are inhibited following stimulation in the PAG or NRM. They are also inhibited by stimulation in the ventral posterior lateral nucleus of the thalamus, sensorimotor cerebral cortex, and reticular formation. Some STT cells can be excited by stimulation in the sensorimotor cortex or in the reticular formation.
19. The inhibitory pathway from the NRM to STT cells descends in the dorsal part of the lateral funiculus, includes myelinated axons, and preferentially inhibits responses to fine fiber inputs as opposed to $A\alpha\beta$ inputs. However, the responses of STT cells to innocuous cutaneous stimuli are usually inhibited to the same extent as are the responses to noxious input. Nucleus raphe magnus

stimulation can produce actions on presynaptic terminals, as well as postsynaptic inhibition of STT cells.

20. Nucleus raphe magnus inhibition of STT cells may involve release of serotonin. Spinothalamic tract cells are inhibited by iontophoretically applied serotonin, and NRM inhibition can be partially blocked by serotonin receptor blockers. However, the receptor blockers are able to reverse PAG inhibition better than NRM inhibition. As in the case of dorsal horn interneurons, the preferential blockade of PAG as opposed to NRM inhibition by serotonin receptor blockers may indicate a serotonergic synapse within the brain stem in the descending pathway from the PAG.
21. Spinothalamic tract cells that project to the medial thalamus behave like reticular formation neurons in having diffuse excitatory receptive fields over much of the body surface. Much of the excitatory input to these cells depend upon a loop through the brain stem, and medially projecting STT cells can be powerfully excited by stimulation within the reticular formation. These cells may participate in the motivational-affective component of pain.
22. Innocuous mechanical stimuli can occasionally inhibit STT cells. Electrical stimulation of axons in the dorsal column is effective in inhibiting STT cells through the action of the volley descending to the segmental level of the STT cell.
23. Spinothalamic tract cells can often be inhibited by noxious stimulation of much of the body surface, a phenomenon that resembles DNIC in dorsal horn interneurons. However, the widespread inhibition of STT cells is in part mediated through segmental pathways, and high-threshold as well as wide dynamic range STT cells are affected.
24. The spinocervical tract may play an important role in nociception, especially in animals like the cat. The spinocervical tract is subject to tonic descending inhibition in decerebrate animals, and there are inhibitory effects when axons in several parts of the spinal cord white matter are stimulated and also when the brain stem is stimulated.
25. It is likely that similar inhibitory controls are exerted on other somatosensory pathways that may be involved in nociception, including the spinoreticular tract and the second-order dorsal column pathway.

6 Correlations Between the Descending Control of Spinal Cord Nociceptive Pathways and the Operation of the Analgesia Systems

6.1 Flexion Reflex

Since analgesia in animals is often tested by flexion reflex responses (such as the tail flick and the jaw-opening reflex), studies concerned with the descending inhibitory control of flexion reflex pathways should correlate well with studies of the intrinsic analgesia systems. In general terms, this is the case, but there have been a number of methodological differences between the two kinds of studies that make a close correlation difficult as yet.

Behavioral studies of analgesia make use of unanesthetized, intact animals. The flexion reflexes tested are often threshold responses to noxious stimuli. In fact, in some cases, such as the jaw-opening reflex, threshold for the reflex appears to be below threshold for pain, and the reflex is more closely related to chewing than to nociception (Dubner et al. 1978; McGrath et al. 1981). The constraint of using awake, behaving animals limits the assessment of pain responses to the lower part of the range of possible stimulus intensities. These factors could lead to misinterpretations if the control of near-threshold responses is qualitatively different from the control of responses to intense stimuli.

On the other hand, studies of the flexion reflex in neurophysiological experiments do allow a consideration of the full range of noxious stimulus intensities, since the animals are protected from pain by decerebration or other lesions, or by anesthesia. However, it is difficult to make a meaningful judgement about the behavioral role of a response in a reduced preparation. For instance, many studies of the flexion reflex have been concerned primarily with the short-latency component of the flexion reflex response that can be recorded from the ventral root in response to stimulation even of large myelinated afferent fibers (e.g., Andén et al. 1966b; Anderson and Shibuya 1966; Lloyd 1943c). However, the component of the flexion reflex that is most likely to account for the long-lasting flexor withdrawal reflex is the late discharge evoked by small afferent fibers (Brooks and Fuortes 1952). This part of the flexion reflex has only occasionally been described in reflex studies (e.g., Laporte and Bessou 1958; LaPorte and Boër 1954, 1955), and little work has been done on the descending control of the portion of the flexion reflex due to C fiber input.

Despite these reservations, a picture has been emerging over the past several decades of an inhibitory control system originating in the brain stem that on the one hand causes reduction in behavioral responses to noxious stimuli and on the other hand an inhibition of the flexion reflex. That these changes can be related to analgesia is suggested by the clinical observation that brain stem or diencephalic stimulation in chronic pain patients can halt or reduce the pain (Hosobuchi et al. 1977a; Richardson and Akil 1977a,b). Thus, although there can be a dissociation between the flexion reflex and pain (McGrath et al. 1981; Willer et al. 1979), studies of the descending control of the flexion reflex can lead to practical applications directed at pain relief.

Curiously, there has been no attempt to examine the neurophysiological basis of the tail flick response and its descending control. This would seem to be an attractive system for a detailed investigation, since the tail flick is so widely used in behavioral studies of analgesia.

Besides the differences in the responses used to test for "analgesia" and for descending inhibition in behavioral and neurophysiological experiments utilizing the flexion reflex, there are also important differences in the means used to activate the descending pathways. The most direct comparison is between the stimuli used for stimulation-produced analgesia (SPA) and for inhibition of the flexion reflex pathway. In the behavioral studies, SPA has been evoked by prolonged trains of stimuli using high stimulus intensities. Neurophysiological experiments, on the other hand, typically employ brief trains of stimuli, often at high frequencies, and low stimulus intensities. The problem in making correlations between the two kinds of experiment is compounded by these differences in stimulus parameters. For example, prolonged trains of high-intensity stimuli may activate synaptically interconnected descending systems by temporal and spatial summation which are not brought into action by the weaker stimuli used in neurophysiological investigations. Thus, it may be difficult to be sure that stimulation in a given brain stem nucleus causes its behavioral effects by way of the same pathway or set of pathways that may be responsible for inhibition in neurophysiological experiments. On the other hand, tissue damage due to chronically implanted electrodes may reduce the effectiveness of electrical stimuli in awake, behaving animals (Zorman et al. 1981). Thus the differences in effective stimulus intensities in the behavioral and electrophysiological experiments may be more apparent than real. It can be concluded that there is convergent evidence from behavioral and neurophysiological studies that important systems exist in the brain stem, particularly near the midline, that result in "analgesia" and inhibition of the flexion reflex. However, the mapping of sites for SPA is more complete than that for inhibition of the flexion reflex.

Pharmacological studies have been done both of the analgesia systems and the pathways controlling the flexion reflex as studied neurophysiologically. An important correspondence has been shown between opiate analgesia and the changes in behavioral tests of "analgesia" (e.g., Herz et al. 1970; Yaksh et al. 1976b). However, little is known about the effects of opiates on the flexion reflex as studied neurophysiologically. This statement is especially true for the late component of the flexion reflex that is evoked by Aδ and C fibers (however, see Bell and Martin 1977).

Related to opiate analgesia are acupuncture, one form of analgesia due to transcutaneous nerve stimulation, and the analgesia due to vaginal stimulation (Hill and Ayliffe 1981; Mayer et al. 1977; Sjölund and Eriksson 1976, 1979). Some work has been done in which the flexion reflex has been used to study the mechanism of acupuncture or of transcutaneous nerve stimulation (Du and Chao 1976). However, more experiments should be done using the late component of the flexion reflex.

Drug studies have implicated central monoaminergic pathways in SPA and in opiate analgesia (Akil and Liebeskind 1975; Vogt 1974). Unfortunately, the evidence with respect to opiate analgesia is not very consistent, perhaps because of the different testing methods used in different laboratories or because of the action at multiple sites in the CNS of drugs given systemically. At any rate, monoamine pathways have also been implicated in the brain stem control of the flexion reflex pathway (Engberg et al. 1968a,b). Again, more work is needed on the role of monoamine pathways in controlling the late component of the flexion reflex.

The pharmacological complexity of the descending control systems is barely surfacing as the role of the endogenous opioid substances and of the monoaminergic pathways becomes better known. There is no reason to believe that a variety of other classical transmitters (e.g., acetylcholine, γ-aminobutyric acid, glycine) and as yet nonclassical transmitters or modulators (substance P, somatostatin, neurotensin, etc.) will be exempt from playing a role in the regulation of nociceptive transmission in the spinal cord.

6.2 Nociceptive Dorsal Horn Neurons

Many of the difficulties in correlating the analgesia mechanisms with descending control of the flexion reflex apply also to the descending control of nociceptive dorsal horn interneurons. For example, it is unclear to what extent the responses of dorsal horn interneurons reflect neural activity directly relatable to pain. Responses of motoneurons in the flexion reflex may not necessarily indicate pain, and responses of interneurons whose synaptic connections are undefined are another step removed from a clear relationship to pain. To some extent, this criticism can be avoided by studying tract cells that provide the brain with the information required for pain sensation. However, even for STT cells, which are very likely to play such a role, our understanding of their precise contribution to pain is still known only in bare outline. The possible contribution of other tracts, such as the spinoreticular, spinocervical, and second-order dorsal column pathways, is quite unclear.

The goal of many of the experiments on descending inhibition of dorsal horn interneurons is to provide a neural correlate of analgesia produced by comparable stimulation in awake, behaving animals. However, an exact correspondence is too much to expect. For example, a given region of the brain may contain elements of more than one system that can be stimulated by electrodes placed in the vicinity. The behavioral effect may be mixed. For example, stimulation in the midbrain can produce aversive reactions during stimulation, but analgesia immediately afterwards (Mayer and Liebeskind 1974). Similar stimulation, however, might be purely inhibitory when assayed by recordings from a particular dorsal horn interneuron (e.g., Carstens et al. 1981a). The interneuron may be providing an appropriate response to match the analgesia, but other neurons may also be engaged that participate in the aversive activity. The same observations can be made for stimulation in the medullary reticular formation (Casey 1971; Haber et al. 1980). It should not be concluded that a particular neuron under observation must necessarily participate in the overt behavior resulting from a given stimulus.

It would be particularly helpful if techniques could be developed to monitor the activity of identified dorsal horn neurons, such as STT cells, in awake, behaving animals. Studies of this kind are currently being done on the trigeminal system by

Dubner's group (Dubner et al. 1981), but comparable work needs to be done on spinal cord neurons.

Another approach that would add to our understanding of pain transmission would be a consideration of the response of the population of nociceptive neurons, along the lines already initiated for primary afferent neurons responsive to tactile and thermal stimuli (Johnson 1974; Johnson et al. 1973).

Although there are at least two different classes of nociceptive dorsal horn cells – high-threshold and wide dynamic range – it is uncertain whether these play similar or different roles in pain sensation. It is easy to see how high-threshold cells might serve in a sensory channel for pain sensation, but it is more difficult to see how the brain can sort out information carried by wide dynamic range cells into, say, tactile messages and pain messages. A case has been made that wide dynamic range neurons are sufficient (Mayer et al. 1975; Price and Mayer 1975) or even crucial (LeBars et al. 1979a,b) for pain. Furthermore, axons in the ventral quadrant of the cord, presumably belonging to the STT, convey surprisingly accurate tactile information (Noordenbos and Wall 1976). Given the small numbers of low-threshold spinothalamic neurons (Chung et al. 1979; Giesler et al. 1981b), the question must be asked whether tactile sensation is a function of wide dynamic range spinothalamic cells. The descending control systems could provide a way to switch wide dynamic range cells from being chiefly responsive to tactile stimuli (if descending inhibition of nociceptive responses were selectively produced; cf., Carstens et al. 1981a; Gerhart et al. 1981a; Liebeskind et al. 1973) or to noxious stimuli (if descending inhibition were just of tactile responses, cf. Coulter et al. 1974). However, there would still need to be an appropriate means for recognition of tactile versus nociceptive information at the level of the brain. A mechanism of this kind could be described as a form of selective attention.

In more general terms, one can speculate that the descending control systems help regulate the activity of the somatosensory pathways in order to make the responsiveness of the pathways appropriate to the behavioral context in which they operate (e.g., Dubner et al. 1981). Thus one might expect, for example, an STT cell to behave quite differently under such different circumstances as during sleep, in anticipation of a severely painful stimulus, or during an escape maneuver.

Although there is little question that there is more than one descending analgesia system, it will be of interest to learn how many analgesia systems there are. For example, will it turn out that there is one system activated by: (a) stimulation in certain loci near the midline of the brain stem; (b) exogenous opiates; (c) acupuncture; (d) low-frequency, high-intensity peripheral nerve stimulation; and (e) vaginal stimulation, and another system activated by: (a) stimulation in the ventrobasal thalamus or internal capsule; (b) nonopiate analgesics; (c) dorsal column stimulation; and (d) high-frequency, low-intensity peripheral nerve stimulation? Does stress-induced analgesia use these analgesia mechanisms or others?

Once we understand the physiology and pharmacology of the descending analgesia systems, it is likely that rationales will develop for their manipulation. For example, it has already been reported that when tolerance to SPA develops, the tolerance can be counteracted by administration of a serotonin precursor (Hosobuchi et al. 1980; Oliveras et al. 1978). Similarly, the effectiveness of acupuncture can be improved by administration of D-amino acids, which are thought to protect the endorphins from enzymatic destruction (Cheng and Pomeranz 1980; Ehrenpreis et al. 1979; Han et al. 1980). These and other similar approaches should in future greatly enhance the clinical possibilities for the treatment of pain.

7 References

Abols IA, Basbaum AI (1981) Afferent connections of the rostral medulla of the cat: a neural substrate for midbrain-medullary interactions in the modulation of pain. J Comp Neurol 201:285–297

Abzug C, Maeda M, Peterson BW, Wilson VJ (1974) (with an appendix by CP Bean) Cervical branching of lumbar vestibulospinal axons. J Physiol (Lond) 243:499–522

Acupuncture Anesthesia (1975) (a translation of a Chinese publication of the same title) US Department of Health, Education and Welfare Publication Number (NIH) 75-784

Adams JE (1976) Naloxone reversal of analgesia produced by brain stimulation in the human. Pain 2:161–166

Adams JE, Hosobuchi Y, Fields HL (1974) Stimulation of internal capsule for relief of chronic pain. J Neurosurg 41:740–744

Adler M, Kostowski W, Recchia M, Samanin R (1975) Anatomical specificity as the critical determinant of the interrelationship between raphe lesions and morphine analgesia. Eur J Pharmacol 32:39–44

Agnew DC, Merskey H (1976) Words of chronic pain. Pain 2:73–81

Akaike A, Shibata T, Satoh M, Takagi H (1978) Analgesia induced by microinjection of morphine into, and electrical stimulation of, the nucleus reticularis paragigantocellularis of rat medulla oblongata. Neuropharmacology 17:775–778

Akil H, Liebeskind JC (1975) Monoaminergic mechanisms of stimulation-produced analgesia. Brain Res 94:279–296

Akil H, Mayer DJ (1972) Antagonism of stimulation-produced analgesia by *p*-CPA, a serotonin synthesis inhibitor. Brain Res 44:692–697

Akil H, Mayer DJ, Liebeskind JC (1976) Antagonism of stimulation-produced analgesia by naloxone, a narcotic antagonist. Science 191:961–962

Akil H, Richardson DE, Barchas JD (1979) Pain control by focal brain stimulation in man: relationship to enkephalins and endorphins. In: Beers RF, Bassett EG (eds) Mechanisms of pain and analgesic compounds, 11th Miles Int Symp. Raven Press, New York

Akil H, Richardson DE, Hughes J, Barchas JE (1978a) Enkephalin-like material elevated in ventricular cerebrospinal fluid of pain patients after analgesic focal stimulation. Science 201:463–465

Akil H, Richardson DE, Barchas JD, Li CH (1978b) Appearance of β-endorphin-like immunoreactivity in human ventricular cerebrospinal fluid upon analgetic electrical stimulation. Proc Natl Acad Sci USA 75:5170–5172

Albe-Fessard D, Levante A, Lamour Y (1974) Origin of spino-thalamic tract in monkeys. Brain Res 65:503–509

Alexandrowicz JS (1951) Muscle receptor organs in the abdomen of *Homarus vulgaris* and *Palinurus vulgaris*. Q J Microscop Sci 92:163–202

Almay BGL, Johansson F, Von Knorring L, Terenius L, Wahlström A (1978) Endorphins in chronic pain. I. Differences in CSF endorphin levels between organic and psychogenic pain syndromes. Pain 5:153–162

Alvord EC, Fuortes MGF (1954) A comparison of flexor reflexes of cutaneous and muscular origin. J Physiol (Lond) 123:251–259

American Acupuncture Anesthesia Study Group (1976) Acupuncture anesthesia in the People's Republic of China. National Academy of Science USA, Washington DC

Amir S, Amit Z (1978) Endogenous opioid ligands may mediate stress-induced changes in the affective properties of pain related behavior in rats. Life Sci 23:1143–1151

Amir S, Amit Z (1979) The pituitary gland mediates acute and chronic pain responsiveness in stressed and non-stressed rats. Life Sci 24:439–448

Andén NE, Häggendal J, Magnusson T, Rosengren E (1964a) The time course of the disappearance of noradrenaline and 5-hydroxytryptamine in the spinal cord after transection. Acta Physiol Scand 62:115–118

Andén NE, Jukes MGM, Lundberg A (1964b) Spinal reflexes and monoamine liberation. Nature 202:1222–1223

Andén NE, Jukes MGM, Lundberg A, Vyklický L (1964c) A new spinal flexor reflex. Nature 202:1344–1345

Andén NE, Jukes MGM, Lundberg A (1966a) The effect of DOPA on the spinal cord. 2. A pharmacological analysis. Acta Physiol Scand 67:387–397

Andén NE, Jukes MGM, Lundberg A, Vyklický L (1966b) The effect of DOPA on the spinal cord. 1. Influence on transmission from primary afferents. Acta Physiol Scand 67:373–386

Andén NE, Jukes MGM, Lundberg A, Vyklický L (1966c) The effect of DOPA on the spinal cord. 3. Depolarization evoked in the central terminals of ipsilateral Ia afferents by volleys in the flexor reflex afferents. Acta Physiol Scand 68:322–336

Andersen P, Eccles JC, Schmidt RF, Yokota T (1964a) Depolarization of presynaptic fibers in the cuneate nucleus. J Neurophysiol 27:92–106

Andersen P, Eccles JC, Schmidt RF, Yokota T (1964b) Identification of relay cells and interneurons in the cuneate nucleus. J Neurophysiol 27:1080–1095

Anderson DG, Jamieson JL, Man SC (1974) Analgesic effects of acupuncture on the pain of ice-water: a double-blind study. Can J Psychol 28:239–244

Anderson EG, Holgerson LO (1966) The distribution of 5-hydroxytryptamine and norepinephrine in cat spinal cord. J Neurochem 13:479–485

Anderson EG, Shibuya T (1966) The effects of 5-hydroxytryptophan and L-tryptophan on spinal synaptic activity. J Pharmacol Exp Ther 153:352–360

Anderson SD, Basbaum AI, Fields HL (1977) Response of medullary raphe neurons to peripheral stimulation and to systemic opiates. Brain Res 123:363–368

Andersson SA, Ericson T, Holmgren E, Lindqvist G (1973) Electro-acupuncture effect on pain threshold measured with electrical stimulation of teeth. Brain Res 63:393–396

Andersson SA, Holmgren E (1975) On acupuncture analgesia and the mechanism of pain. Am J Chin Med 3:311–334

Andrews HC, Workman W (1941) Pain threshold measurements in the dog. J Pharmacol Exp Ther 73:99–103

Angaut-Petit D (1975) The dorsal column system: II. Functional properties and bulbar relay of the postsynaptic fibres of the cat's fasciculus gracilis. Exp Brain Res 22:471–493

Applebaum AE, Beall JE, Foreman RD, Willis WD (1975) Organization and receptive fields of primate spinothalamic tract neurons. J Neurophysiol 38:572–586

Armett CJ, Gray JAB, Palmer JF (1961) A group of neurones in the dorsal horn associated with cutaneous mechanoreceptors. J Physiol 156:611–622

Aronin N, DiFiglia M, Liotta AS, Martin JB (1981) Ultrastructural localization and biochemical features of immunoreactive leu-enkephalin in monkey dorsal horn. J Neurosci 1:561–577

Ashkenazi R, Haber B (1975) Mode of action of a sympathomimetic amine, paramethoxyphenylethylamine (PMPEA) in the mouse CNS: release of biogenic amines. Neurosci Lett 1:163–168

Atweh SF, Kuhar MJ (1977a) Autoradiographic localization of opiate receptors in rat brain. I. Spinal cord and lower medulla. Brain Res 124:53–67

Atweh SF, Kuhar MJ (1977b) Autoradiographic localization of opiate receptors in rat brain. II. The brain stem. Brain Res 129:1–12

Atweh SF, Kuhar MJ (1977c) Autoradiographic localization of opiate receptors in rat brain. III. The telencephalon. Brain Res 134:393–405
Audigier Y, Mazarguil H, Gout R, Cros J (1980) Structure-activity relationships of enkephalin analogs at opiate and enkephalin receptors: correlation with analgesia. Eur J Pharmacol 63:35–46
Augustinsson LE, Bohlin P, Bundsen P, Carlsson CA, Forssman L, Sjöberg P, Tyreman No (1977) Pain relief during delivery by transcutaneous electrical nerve stimulation. Pain 4:59–65
Austin GM (1952) Suprabulbar mechanisms of facilitation and inhibition of cord reflexes. Res Publ Assoc Nerv Ment Dis 30:192–222
Ayhan IH (1972) Effect of 6-hydroxydopamine on morphine analgesia. Psychopharmacologia 25:183–188
Baizman ER, Cox BM, Osman OH, Goldstein A (1979) Experimental alterations of endorphin levels in rat pituitary. Neuroendocrinology 28:402–414
Bajusz S, Rónai AZ, Székely JI, Dunai-Kovács Z, Berzétei I, Gráf, L (1976) Enkephalin analogs with enhanced opiate activity. Acta Biochem Biophys Acad Sci Hung 11:305–309
Baker RG, Anderson EG (1970) The effects of L-3,4-dihydroxyphenylalanine on spinal reflex activity. J Pharm Exp Ther 173:212–223
Balagura S, Ralph T (1973) The analgesic effect of electrical stimulation of the diencephalon and mesencephalon. Brain Res 60:369–379
Barber J, Mayer DJ (1977) Evaluation of the efficacy and neural mechanism of a hypnotic analgesia procedure in experimental and clinical dental pain. Pain 4:41–48
Barton C, Basbaum AI, Fields HL (1980) Dissociation of supraspinal and spinal actions of morphine: a quantitative evaluation. Brain Res 188:487–498
Basbaum AI (1981) Descending control of pain transmission: possible serotonergic-enkephalinergic interactions. Adv Exp Med Biol 133:177–189
Basbaum AI, Fields HL (1978) Endogenous pain control mechanisms: Review and hypothesis. Ann Neurol 4:451–462
Basbaum AI, Fields HL (1979) The origin of descending pathways in the dorsolateral funiculus of the spinal cord of the cat and rat: further studies on the anatomy of pain modulation. J Comp Neurol 187:513–532
Basbaum AI, Clanton CH, Fields HL (1976) Opiate and stimulus-produced analgesia: functional anatomy of a medullospinal pathway. Proc Natl Acad Sci USA 73:4685–4688
Basbaum AI, Clanton CH, Fields HL (1978) Three bulbospinal pathways from the rostral medulla of the cat: an autoradiographic study of pain modulating systems. J Comp Neurol 178:209–224
Basbaum AI, Marley NJE, O'Keefe JO, Clanton CH (1977) Reversal of morphine and stimulus-produced analgesia by subtotal spinal cord lesions. Pain 3:43–56
Battista AF, Wolff BB (1973) Levodopa and induced-pain response. A study of patients with Parkinsonian and pain syndromes. Arch Intern Med 132:70–74
Baxendale RH, Ferrell WR (1981) The effect of knee joint afferent discharge on transmission in flexion reflex pathways in decerebrate cats. J Physiol (Lond) 315:231–242
Beall JE, Martin RF, Applebaum AE, Willis WD (1976) Inhibition of primate spinothalamic tract neurons by stimulation in the region of the nucleus raphe magnus. Brain Res 114:328–333
Beall JE, Applebaum AE, Foreman RD, Willis WD (1977) Spinal cord potentials evoked by cutaneous afferents in the monkey. J Neurophysiol 40:199–211
Beaumont A, Hughes J (1979) Biology of opioid peptides. Ann Rev Pharm Tox 19:245–267
Becker DP, Glude H, Nulsen FE, Jane JA (1969) An inquiry into the neurophysiological basis for pain. J Neurosurg 30:1–13

Beecher HK (1959) Measurement of subjective responses. Oxford University Press, New York

Behbehani MM, Fields HL (1979) Evidence that an excitatory connection between the periaqueductal gray and nucleus raphe magnus mediates stimulation produced analgesia. Brain Res 170:85 – 93

Behbehani MM, Pomeroy SL (1978) Effect of morphine injected in the periaqueductal gray on the activity of single units in nucleus raphe magnus of the rat. Brain Res 149:266 – 269

Belcher G, Ryall RW, Schaffner R (1978) The differential effects of 5-hydroxytryptamine, noradrenaline and raphe stimulation on nociceptive and nonnociceptive dorsal horn interneurones in the cat. Brain Res 151:307 – 321

Bell JA, Martin WR (1977) The effect of the narcotic antagonists naloxone, naltrexone and nalorphine on spinal cord C-fiber reflexes evoked by electrical stimulation or radiant heat. Eur J Pharmacol 42:147 – 154

Belluzzi JD, Grant N, Garsky V, Sarantakis D, Wise CD, Stein L (1976) Analgesia induced *in vivo* by central administration of enkephalin in rat. Nature 260:625 – 626

Bennett GJ, Abdelmoumene M, Hayashi H, Dubner R (1980) Physiology and morphology of substantia gelatinosa neurons intracellularly stained with horseradish peroxidase. J Comp Neurol 194:809 – 827

Bennett GJ, Hayashi H, Abdelmoumene M, Dubner R (1979) Physiological properties of stalked cells of the substantia gelatinosa intracellularly stained with horseradish peroxidase. Brain Res 164:285 – 289

Bennett GJ, Mayer DJ (1979) Inhibition of spinal cord interneurons by narcotic microinjection and focal electrical stimulation in the periaqueductal central gray matter. Brain Res 172:243 – 258

Bergland RM, Page RB (1978) Can the pituitary secrete directly to the brain? (Affirmative anatomical evidence) Endrocrinology 102:1325 – 1338

Berkley K (1980) Spatial relationships between the terminations of somatic sensory and motor pathways in the rostral brainstem of cats and monkeys. I. Ascending somatic sensory inputs to lateral diencephalon. J Comp Neurol 193:283 – 317

Berlin FS, Bartlett RL, Black JD (1975) Acupuncture and placebo: effects on delaying the terminating response to a painful stimulus. Anesthesiology 42:527 – 531

Berman AL (1968) The brain stem of the cat. A cytoarchitectonic atlas with stereotaxic coordinates. Univ Wisconsin Press, Madison

Bernston GG, Berson BS (1980) Antinociceptive effects of intraventricular or systemic administration of vasopressin in the rat. Life Sci 26:455 – 459

Besson JM, Conseiller C, Hamann KF, Maillard MC (1972) Modifications of dorsal horn cell activities in the spinal cord, after intra-arterial injection of bradykinin. J Physiol (Lond) 221:189 – 205

Besson JM, Guilbaud G, LeBars D (1975) Descending inhibitory influences exerted by the brain stem upon the activities of dorsal horn lamina V cells induced by intra-arterial injection of bradykinin into the limbs. J Physiol (Lond) 248:725 – 739

Besson JM, Oliveras JL, Chaouch A, Rivot JP (1981) Role of the raphe nuclei in stimulation producing analgesia. Adv Exp Med Biol 133:153 – 176

Bhargara HN, Way EL (1972) Acetylcholinesterase inhibition and morphine effects in morphine tolerant and dependent mice. J Pharmacol Exp Ther 183:31 – 40

Bhattacharya SK, Keshary PR, Sanyal AK (1978) Immobilization stress-induced antinociception in rats: possible role of serotonin and prostaglandins. Eur J Pharmacol 50:83 – 85

Biscoe TJ, Curtis DR, Ryall RW (1966) An investigation of catecholamine receptors of spinal interneurons. Int J Neuropharmacol 5:429 – 434

Bläsig J, Reinhold K, Herz A (1973) Effect of 6-hydroxydopamine, 5,6-dihydroxytryptamine and and raphe lesions on the antinociceptive actions of morphine in rats. Psychopharmacologia 31:111 – 119

Blessing WW, Chalmers JP (1979) Direct projection of catecholamine (presumably dopamine)-containing neurons from hypothalamus to spinal cord. Neurosci Lett 11:35 – 40

Bloom F, Battenberg E, Rossier J, Ling N, Guillemin R (1978) Neurons containing β-endorphin in rat brain exist separately from those containing enkephalin: immunocytochemical studies. Proc Natl Acad Sci USA 75:1591 – 1595

Bobillier P, Seguin S, Petitjean F, Salvert D, Touret M, Jouvet M (1976) The raphe nuclei of the cat brain stem: a topographical atlas of their efferent projections as revealed by autoradiography. Brain Res 113:449 – 486

Bodnar RJ, Kelly DD, Glusman M (1978a) Stress-induced analgesia: Time course of pain reflex alterations following cold water swims. Bull Psychonom Soc 11:333 – 336

Bodnar RJ, Kelly DD, Spiaggia A, Ehrenberg C, Glusman M (1978b) Dose-dependent reductions by naloxone of analgesia induced by cold-water stress. Pharmcol Biochem Behav 8:667 – 672

Bodnar RJ, Kelly DD, Spiaggia A, Glusman M (1978c) Stress-induced analgesia: Adaptation following chronic cold water swims. Bull Psychonom Soc 11:337 – 340

Bodnar RJ, Kelly DD, Steiner SS, Glusman M (1978d) Stress-induced analgesia and morphine-produced analgesia: lack of cross-tolerance. Pharmacol Biochem Behav 8:661 – 666

Bodnar RJ, Glusman M, Brutus M, Spiaggia A, Kelly DD (1979a) Analgesia induced by cold-water stress: attenuation following hypophysectoma. Physiol Behav 23:53 – 62

Bodnar RJ, Kelly DD, Glusman M (1979b) 2-Deoxy-D-glucose analgesia: influences of opiate and non-opiate factors. Pharmacol Biochem Behav 11:297 – 301

Bodnar RJ, Kelly DD, Mansour A, Glusman M (1979c) Differential effects of hypophysectomy upon analgesia induced by two glucoprivic stressors and morphine. Pharmacol Biochem Behav 11:303 – 308

Bodnar RJ, Kelly DD, Brutus M, Glusman M (1980) Stress-induced analgesia: neural and hormonal determinants. Neurosci Biobehav Rev 4:87 – 100

Boëthius J, Lindblom U, Meyerson BA, Widén L (1976) Effects of multifocal brain stimulation on pain and somatosensory functions. In Zottermann Y (ed) Sensory functions of the skin in primates. Pergamon, Oxford, pp 531 – 546

Boivie J (1979) An anatomical reinvestigation of the termination of the spinothalamic tract in the monkey. J Comp Neurol 186:343 – 370

Boivie JJG, Perl ER (1975) Neural substrates of somatic sensation. In Hunt CC (ed) MTP International review of science, Physiology series one, vol 3, Neurophysiology. University Park Press, Baltimore, pp 303 – 411

Bonica JJ (1974) Therapeutic acupuncture in the People's Republic of China. Implications for American medicine. JAMA 228:1544 – 1551

Bonnycastle DD (1962) The use of animals in the study of analgetic drugs. In Keele CA, Smith R (eds) The assessment of pain in man and animals. Livingstone, London, pp 231 – 243

Boren JJ, Malis JL (1961) Determining theresholds of aversive brain stimulation. Am J Physiol 201:429 – 433

Bourgoin S, Oliveras JL, Bruxelle J, Hamon M, Besson JM (1980) Electrical stimulation of the nucleus raphe magnus in the rat. Effects on 5-HT metabolism in the spinal cord. Brain Res 194:377 – 389

Bowd AD (1980) Ethics and animal experimentation. Am Psychol 35:224 – 225

Bowker RM, Steinbusch HWM, Coulter JD (1981) Serotonergic and peptidergic projections to the spinal cord demonstrated by a combined retrograde HRP histochemical and immunocytochemical staining method. Brain Res 211:412 – 417

Bowker RM, Westlund KN, Coulter JD (1981) Origins of serotonergic projections to the spinal cord in rat: an immunocytochemical-retrograde transport study. Brain Res 226:181 – 199

Bowsher D (1961) The termination of secondary somatosensory neurons within the thalamus of *Macaca mulatta:* an experimental degeneration study. J Comp Neurol 117:213 – 227

Bowsher D (1976) Role of the reticular formation in responses to noxious stimulation. Pain 2:361 – 378

Bradbury AG, Smyth DG, Snell CR, Birdsall NJM, Hulme EC (1976) C fragment of lipotropin has a high affinity for brain opiate receptors. Nature 260:793 – 795

Bradley PB, Wolstencroft JH (1965) Actions of drugs on single neurones in the brain stem. Br Med Bull 21:15 – 18

Breglio V, Anderson DC, Merrill HK (1970) Alteration in footshock threshold by low-level septal brain stimulation. Physiol Behav 5:715 – 719

Brennan RW, Veldhuis J, Chu R (1973) Acupuncture anesthesia. Lancet 2:849 – 850

Briggs I (1977) Excitatory responses of neurones in rat bulbar reticular formation to bulbar raphe stimulation and to iontophoretically applied 5-hydroxytryptamine, and their blockade by LSD 25. J Physiol (Lond) 265:327 – 340

Brodal A, Taber E, Walberg F (1960) The raphe nuclei of the brain stem in the cat. II. Efferent connections. J Comp Neurol 114:239 – 260

Brooks C McC, Fuortes MGF (1952) The relation of dorsal and ventral root potentials to reflex activity in mammals. J Physiol (Lond) 116:380 – 394

Brouwer B (1933) Centrifugal influence on centripetal systems in the brain. J Nerve Ment Dis 77:621 – 627

Brown AG (1971) Effects of descending impulses on transmission through the spinocervical tract. J Physiol 219:103 – 125

Brown AG, Franz DN (1969) Responses of spinocervical tract neurones to natural stimulation of identified cutaneous receptors. Exp Brain Res 7:231 – 249

Brown AG, Kirk EJ, Martin HF (1973) Descending and segmental inhibition of transmission through the spinocervical tract. J Physiol (Lond) 230:689 – 705

Brown AG, Coulter JD, Rose PK, Short AD, Snow PJ (1977) Inhibition of spinocervical tract discharges from localized areas of the sensorimotor cortex in the cat. J Physiol (Lond) 264:1 – 16

Brown TG (1911) The intrinsic factors in the act of progression in the mammal. Proc R Soc Lond [Biol] 84:308 – 319

Brown TG (1912) The factors in rhythmic activity of the nervous system. Proc R Soc Lond [Biol] 85:278 – 289

Brown TG (1914) On the nature of the fundamental activity of the nervous centres; together with an analysis of rhythmic activity in progression, and a theory of the evolution of function in the nervous system. J Physiol (Lond) 48:18 – 46

Buchsbaum MS, Davis GC, Bunney WE (1977) Naloxone alters pain perception and somatosensory evoked potentials in normal subjects. Nature 270:620 – 622

Buckett WR (1979) Peripheral stimulation in mice induces short-duration analgesia preventable by naloxone. Eur J Pharmacol 58:169 – 178

Burke RE, Rudomin P (1977) Spinal neurons and synapses. In Kandel ER (ed) Cellular biology of neurons, part 2. Am Physiol Soc, Bethesda, (Handbook of physiology, sect 1, vol 1, pp 877 – 944)

Burke RE, Jankowska E, ten Bruggencate G (1970) A comparison of peripheral and rubrospinal synaptic input to slow and fast twitch motor units of triceps surae. J Physiol (Lond) 207:709 – 732

Burke RE, Rudomín P, Vyklický L, Zajac FE (1971) Primary afferent depolarization and flexion reflexes produced by radiant heat stimulation of the skin. J Physiol (Lond) 213:185 – 214

Buscher HH, Hill RC, Römer D, Cardinaux F, Closse A, Hauser D, Pless J (1976) Evidence for analgesic activity of enkephalin in the mouse. Nature 261:423 – 425

Buxbaum DM, Yarbrough GG, Carter ME (1973) Biogenic amines and narcotic effects. I Modification of morphine-induced analgesia and motor activity after alteration of cerebral amine levels. J Pharmacol Exp Ther 185:317 – 327

Cajal S Ramon Y (1909) Histologie du système nerveux de l'homme et des vertébrés, Vol II, Madrid, Inst Cajal

Calvillo O, Henry JL, Neuman RS (1974) Effects of morphine and naloxone on dorsal horn neurones in the cat. Can J Physiol Pharmacol 52:1207 – 1211

Campbell JN, Taub A (1973) Local analgesia from percutaneous electrical stimulation. Arch Neurol 28:347 – 350

Cangiano A, Lutzemberger L (1972) The action of selectively activated group II muscle afferent fibers on extensor motoneurons. Brain Res 41:475 – 478

Cannon JT, Liebeskind JC (1979) Descending control systems. In Beers, Jr RF, Bassett EG (eds) Mechanisms of pain and analgesic compounds. Raven Press, New York, pp 171 – 184

Carli G, Diete-Spiff K, Pompeiano O (1967a) Transmission of sensory information through the lemniscal pathway during sleep. Arch Ital Biol 105:31 – 51

Carli G, Diete-Spiff K, Pompeiano O (1967b) Presynaptic and postsynaptic inhibition of transmission of somatic afferent volleys through the cuneate nucleus during sleep. Arch Ital Biol 105:52 – 82

Carlsson A, Falck B, Fuxe K, Hillarp NA (1964) Cellular localization of monoamines in the spinal cord. Acta Physiol Scand 60:112 – 119

Carpenter D, Engberg I, Lundberg A (1965) Differential supraspinal control of inhibitory and excitatory actions from the FRA to ascending spinal pathways. Acta Physiol Scand 63:103 – 110

Carpenter D, Engberg I, Lundberg A (1966) Primary afferent depolarization evoked from the brain stem and the cerebellum. Arch Ital Biol 104:73 – 85

Carstens E, Bihl H, Irvine DRF, Zimmermann M (1981a) Descending inhibition from medial and lateral midbrain of spinal dorsal horn neuronal responses to noxious and nonnoxious cutaneous stimuli in the cat. J Neurophysiol 45:1029 – 1042

Carstens E, Fraunhoffer M, Zimmermann M (1981b) Serotenergic mediation of descending inhibition from midbrain periaqueductal gray, but not reticular formation, of spinal nociceptive transmission in the cat. Pain 10:149 – 167

Carstens E, Klumpp D, Zimmermann M (1979a) The opiate antagonist, naloxone, does not affect descending inhibition from midbrain of nociceptive spinal neuronal discharges in the cat. Neurosci Lett 11:323 – 327

Carstens E, Tulloch I, Zieglgänsberger W, Zimmermann M (1979b) Presnaptic excitability changes induced by morphine in single cutaneous afferent C- and A-fibers. Pfluegers Arch 379:143 – 147

Carstens E, Yokota T, Zimmermann M (1979c) Inhibition of spinal neuronal responses to noxious skin heating by stimulation of mesencephalic periaqueductal gray in the cat. J Neurophysiol 42:558 – 568

Carstens E, Klumpp D, Zimmermann M (1980a) Differential inhibitory effects of medial and lateral midbrain stimulation on spinal neuronal discharges to noxious skin heating in the cat. J Neurophysiol 43:332 – 342

Carstens E, Klumpp D, Zimmermann M (1980b) Time course and effective sites for inhibition from midbrain periaqueductal gray of spinal dorsal horn neuronal responses to cutaneous stimuli in the cat. Exp Brain Res 38:425 – 430

Casey KL (1971) Escape elicited by bulboreticular stimulation in the cat. Int J Neurosci 2:29 – 34

Castiglioni AJ, Gallaway MC, Coulter JD (1978) Spinal projections from the midbrain in monkey. J Comp Neurol 178:329 – 346

Cauthen JC, Renner EJ (1975) Transcutaneous and peripheral nerve stimulation for chronic pain states. Surg Neurol 4:102–104

Cervero F, Iggo A (1980) The substantia gelatinosa of the spinal cord; a critical review. Brain 103:717–772

Cervero F, Iggo A, Ogawa H (1976) Nociceptor-driven dorsal horn neurones in the lumbar spinal cord of the cat. Pain 2:5–24

Cervero F, Iggo A, Molony V (1977a) Responses of spinocervical tract neurones to noxious stimulation of the skin. J Physiol (Lond) 267:537–558

Cervero F, Molony V, Iggo A (1977b) Extracellular and intracellular recordings from neurones in the substantia gelatinosa Rolandi. Brain Res 136:565–569

Cervero F, Iggo A, Molony V (1979a) An electrophysiological study of neurones in the substantia gelatinosa Rolandi of the cat's spinal cord. Q J Exp Physiol 64:297–314

Cervero F, Molony V, Iggo A (1979b) Supraspinal linkage of substantia gelatinosa neurones: effects of descending impulses. Brain Res 175:351–355

Cesa-Bianchi MG, Mancia M, Sotgiu ML (1968) Depolarization of afferent fibers to the Goll and Burdach nuclei induced by stimulation of the brain-stem. Exp Brain Res 5:1–15

Cesa-Bianchi MG, Sotgiu ML (1969) Control by brain stem reticular formation of sensory transmission in Burdach nucleus. Analysis of single units. Brain Res 13:129–139

Chambers WW, Liu CN (1957) Cortico-spinal tract of the cat. An attempt to correlate the pattern of degeneration with deficits in reflex activity following neocortical lesions. J Comp Neurol 108:23–56

Chambers WW, Liu CN, McCouch GP (1970) Cutaneous reflexes and pathways affecting them in the monkey, *Macaca mulatta*. Exp Neurol 28:243–256

Chance WT, Rosecrans JA (1979) Lack of effect of naloxone on autoanalgesia. Pharmac Biochem Behav 11:643–646

Chang HT (1979) Acupuncture analgesia today. Chin Med J [Engl] 92:7–16

Chang KJ, Cuatrecasas P (1979) Multiple opiate receptors: enkephalins and morphine bind to receptors of different specificity. J Biol Chem 254:2610–2618

Chang KJ, Cooper BR, Hazum E, Cuatrecasas P (1979) Multiple opiate receptors: different regional distribution in the brain and differential binding of opiates and opioid peptides. Mol Pharmacol 16:91–104

Chan-Palay V (1979) Combined immunocytochemistry and autoradiography after *in vivo* injections of monoclonal antibody to substance P and ^{3}H-serotonin: co-existence of two putative transmitters in single raphe cells and fiber plexuses. Anat Embryol 156:241–254

Chapman CR (1976) Signal detection theory and pain. Anesthesiology 44:358–359

Chapman CR (1977) Sensory decision theory methods in pain research: a reply to Rollman. Pain 3:295–305

Chapman CR, Benedetti C (1977) Analgesia following transcutaneous electrical stimulation and its partial reversal by a narcotic antagonist. Life Sci 21:1645–1648

Chapman CR, Gehrig JD, Wilson ME (1975a) Acupuncture compared with 33 per cent nitrous oxide for dental analgesia: a sensory decision theory evaluation. Anesthesiology 42:532–537

Chapman CR, Gehrig JD, Wilson ME (1975b) Acupuncture, pain, and signal detection theory. Science 189:65

Chapman CR, Wilson ME, Gehrig JD (1976) Comparative effects of acupuncture and transcutaneous stimulation on the perception of painful dental stimuli. Pain 2:265–283

Chapman CR, Chen AC, Bonica JJ (1977) Effects of intrasegmental electrical acupuncture on dental pain: evaluation by threshold estimation and sensory decision theory. Pain 3:213–227

Chapman CR, Colpitts YM, Benedetti C, Kitaeff R, Gehrig JD (1980) Evoked potential assessment of acupunctural analgesia: attempted reversal with naloxone. Pain 9:183–197

Chapman WP, Jones CM (1944) Variations in cutaneous and visceral pain sensitivity in normal subjects. J Clin Invest 23:81 – 91
Cheney DL, Goldstein A (1971) The effect of *p*-chlorophenylalanine on opiate-induced running, analgesia, tolerance and physical dependence in mice. J Pharmacol Exp Ther 177:309 – 315
Cheng RSS, Pomeranz B (1979) Electroacupuncture analgesia could be mediated by at least two pain-relieving mechanisms; endorphin and non-endorphin systems. Life Sci 25:1957 – 1962
Cheng RSS, Pomeranz B (1980) A combined treatment with D-amino acids and electroacupuncture produces a greater analgesia than either treatment alone; naloxone reverses these effects. Pain 8:231 – 236
Cheng RSS, Pomeranz B (1981) Monoaminergic mechanism of electroacupuncture analgesia. Brain Res 215:77 – 92
Cheung AL, Goldstein A (1976) Failure of hypophysectomy to alter brain content of opioid peptides (endorphins). Life Sci 19:1005 – 1008
Chiang C, Chang C, Chu H, Yang L (1973) Peripheral afferent pathway for acupuncture analgesia. Sci Sin 16:210 – 217
Chiang CY, Liu JY, Chu TH, Pai YH, Chang SC (1975) Studies on spinal ascending pathway for effect of acupuncture analgesia in rabbits. Sci Sin 18:651 – 658
Chiang CY, Tu HC, Chao YF, Pai YH, Ku HK, Cheng JK, Shan HY, Yang FY (1979) Effect of electrolytic lesions or intracerebral injections of 5,6-dihydroxytryptamine in raphe nuclei on acupuncture analgesia in rats. Chin Med J [Engl] 92:129 – 136
Christensen BN, Perl ER (1970) Spinal neurons specifically excited by noxious or thermal stimuli: marginal zone of the dorsal horn. J Neurophysiol 33:293 – 307
Chung JM, Kenshalo DR, Jr, Gerhart KD, Willis WD (1979) Excitation of primate spinothalamic neurons by cutaneous C-fiber volleys. J Neurophysiol 42:1354 – 1369
Cicero TJ, Meyer ER, Smithloff BR (1974) Alpha adrenergic blocking agents: antinociceptive activity and enhancement of morphine-induced analgesia. J Pharmacol Exp Ther 189:72 – 82
Clark WC (1969) Sensory-decision theory analysis of the placebo effect on the criterion for pain and thermal sensitivity. J Abnorm Psychol 74:363 – 371
Clark WC (1974) Pain sensitivity and the report of pain. Anesthesiology 40:272 – 287
Clark WC, Mehl L (1971) A sensory decision theory analysis of the effect of age and sex on d's, various response criteria, and 50% pain threshold. J Abnorm Psychol 78:202 – 212
Clark WC, Yang JC (1974) Acupunctural analgesia? Evaluation by signal detection theory. Science 184:1096 – 1098
Clineschmidt BV, Anderson EG (1970) The blockade of bulbospinal inhibition by 5-hydroxytryptamine antagonists. Exp Brain Res 11:175 – 186
Clineschmidt BV, Pierce JE, Lovenberg W (1971) Tryptophan hydroxylase and serotonin in spinal cord and brain stem before and after chronic transection. J Neurochem 18:1593 – 1596
Collier HOJ (1962) Multiple toe-pinch test for potential analgesic drugs. In Keele CA, Smith R (eds) The assessment of pain in man and animals. Latimer, Trend, Plymouth, pp 262 – 270
Collier HOJ (1972) Pharmacological mechanisms of drug dependence. In Cochin J Pharmacology and the future of man. Karger, Basel 1:65 – 76
Collier HOJ, Schnedier C (1969) Profiles of activity in rodents of some narcotic and narcotic antagonist drugs. Nature 224:610 – 612
Cooper S, Denny-Brown DE, Sherrington C (1927) Interaction between ipsilateral spinal reflexes acting on the flexor muscles of the hind-limb. Proc R Soc Lond [Biol] 101:262 – 303

Cooperman AM, Hall B, Mikalacki K, Hardy R, Sadar E (1977) Use of transcutaneous electrical stimulation in control of postoperative pain – Results of a prospective, randomized, controlled study. Am J Surg 133:185–187

Couch JR (1970) Responses of neurons in the raphe nuclei to serotonin, norepinephrine and acetylcholine and their correlation with an excitatory synaptic input. Brain Res 19:137–150

Coulter JD (1974) Sensory transmission through lemniscal pathway during voluntary movement in the cat. J Neurophysiol 37:831–845

Coulter JD, Maunz RA, Willis WD (1974) Effects of stimulation of sensorimotor cortex on primate spinothalamic neurons. Brain Res 65:351–356

Covino BG, Dubner R, Gybels J, Kosterlitz HW, Liebeskind JC, Sternbach RA, Vyklický L, Yamamura H, Zimmermann M (1980) Ethical standards for investigations of experimental pain in animals. Pain 9:141–143

Cox BM, Goldstein A, Li CH (1976) Opioid activity of a petide, β-lipotropin-(61–91), derived from β-lipotropin. Proc Natl Acad Sci USA 73:1821–1823

Cox BM, Opheim KE, Teschemacher H, Goldstein A (1975) A peptide-like substance from pituitary that acts like morphine. 2. Purification and properties. Life Sci 16:1777–1782

Cox VC, Valenstein ES (1965) Attenuation of aversive properties of peripheral shock by hypothalamic stimulation. Science 149:323–325

Crasilneck HB, Stirman JA, Wilson BJ, McCranie EJ, Fogelman MJ (1955) Use of hypnosis in the management of patients with burns. JAMA 158:103–106

Creed RS, Denny-Brown D, Eccles JC, Liddell EGT, Sherrington CS (1932) Reflex activity of the spinal cord. Oxford Univ Press, London

Creed RS, Sherrington C (1926) Observations on concurrent contraction of flexor muscles in the flexion reflex. Proc Roy Soc Lond [Biol] 100:258–267

Criswell HE (1976) Analgesia and hyperreactivity following morphine microinjection into mouse brain. Pharmacol Biochem Behav 4:23–26

Crowley WR, Jacobs R, Volpe J, Rodriguez-Sierra JF, Komisaruk BR (1976) Analgesic effect of vaginal stimulation in rats: modulation by graded stimulus intensity and hormones. Physiol Behav 16:483–488

Crowley WR, Rodriguez-Sierra JF, Komisaruk BR (1977a) Analgesia induced by vaginal stimulation in rats is apparently independent of a morphine-sensitive process. Psychopharmacology 54:223–225

Crowley WR, Rodriguez-Sierra JF, Komisaruk BR (1977b) Monoaminergic mediation of the antinociceptive effect of vaginal stimulation in rats. Brain Res 137:67–84

Croze S, Antonietti C, Duclaux R (1976) Changes in burning pain threshold induced by acupuncture in man. Brain Res 104:335–340

Crutcher KA, Humbertson AO, Martin GF (1978) The origin of brainstem-spinal pathways in the North American opossum *(Didelphis virginiana)*. Studies using the horseradish peroxidase method. J Comp Neurol 179:169–194

Dahlström A, Fuxe K (1964) Evidence for the existence of monoamine-containing neurons in the central nervous system. I. Demonstration of monoamines in the cell bodies of brain stem neurones. Acta Physiol Scand [Suppl 232] 62:1–55

Dahlström A, Fuxe K (1965) Evidence for the existence of monoamine neurons in the central nervous system. II. Experimentally induced changes in the intraneuronal amine levels of bulbospinal neuron systems. Acta Physiol Scand [Suppl 247] 64:1–36

D'Amour FE, Smith DL (1941) A method for determining loss of pain sensation. J Pharmacol Exp Ther 72:74–79

Dandiya PC, Menon MK (1963) Studies on central nervous system depressants. III. Influence of some tranquillizing agents on morphine analgesia. Arch Int Pharmacodyn Ther 141:223–232

Dawson GD (1958) The central control of sensory inflow. Proc R Soc Med 51:531–535

Day RL, Kitahata LM, Kao FF, Motoyama EK, Hardy JD (1975) Evaluation of acupuncture anesthesia: a psychophysical study. Anesthesiology 43:507–517
Dehen H, Willer JC, Boureau F, Cambier J (1977) Congenital insensitivity to pain, and endogenous morphine-like substances. Lancet 2:293–294
Della Bella D, Casacci F, Sassi A (1978) Opiate receptors: Different ligand affinity in various brain regions. In: Costa E, Trabucchi M (eds) The endorphins. Raven Press, New York, pp 271–277
Dewey WL, Snyder JW, Harris LS, Howes JF (1969) The effects of narcotics and narcotic antagonists on the tail-flick response in spinal mice. J Pharm Pharmacol 21:548–550
Di Carlo V, Hubbard JE, Pate P (1974) Fluorescence histochemistry of monoamine-containing cell bodies in the brain stem of the squirrel monkey (Saimiri sciureus). IV. An atlas. J Comp Neurol 152:347–372
Dickenson AH, Oliveras JL, Besson JM (1979) Role of the nucleus raphe magnus in opiate analgesia as studied by the microinjection technique in the rat. Brain Res 170:95–111
Dickenson AH, Rivot JP, Chaouch A, Besson JM, Le Bars D (1981) Diffuse noxious inhibitory controls (DNIC) in the rat with or without *p*CPA pretreatment. Brain Res 216:313–321
Dilly PN, Wall PD, Webster KE (1968) Cells of origin of the spinothalamic tract in the cat and rat. Exp Neurol 21:550–562
Dorman HL, Gage TW (1978) Effect of electro-acupuncture on the threshold for eliciting the jaw depressor reflex in cats. Arch Oral Biol 23:505–506
Dostrovsky JO, Deakin JFW (1977) Periaqueductal grey lesions reduce morphine analgesia in the rat. Neurosci Lett 4:99–103
Du HJ, Chao YF (1976) Localization of central structures involved in descending inhibitory effect of acupuncture on viscero-somatic reflex discharges. Sci Sin 19:137–148
Dubner R, Sessle BJ, Storey AT (1978) The neural basis of oral and facial function. Plenum Press, New York
Dubner R, Hoffman DS, Hayes RL (1981) Neuronal activity in medullary dorsal horn of awake monkeys trained in a thermal discrimination task. III. Taskrelated responses and their functional role. J Neurophysiol 46:444–464
Dubuisson D, Dennis SG (1977) The formalin test: A quantitative study of the analgesic effects of morphine, meperidine, and brain stimulation in rats and cats. Pain 4:161–174
Dubuisson D, Wall PD (1980) Descending influences on receptive fields and activity of single units recorded in laminae 1, 2 and 3 of cat spinal cord. Brain Res 199:283–298
Dudel J, Kuffler SW (1961) Presynaptic inhibition at the crayfish neuromuscular junction. J Physiol (Lond) 155:543–562
Duggan AW, Griersmith BT (1979) Inhibition of the spinal transmission of nociceptive information by supraspinal stimulation in the cat. Pain 6:149–161
Duggan AW, Hall JG, Headley PM (1976) Morphine, enkephalin and the substantia gelatinosa. Nature 264:456–458
Duggan AW, Griersmith BT, Johnson SM (1981) Supraspinal inhibition of the excitation of dorsal horn neurones by impulses in unmyelinated primary afferents: lack of effect by strychnine and bicuculline. Brain Res 210:231–241
Duncan GH, Gregg JM, Ghia JN (1978) The pain profile: a computerized system for assessment of chronic pain. Pain 5:275–284
Dupont A, Lépine J, Langelier P, Mérand Y, Rouleau D, Vaudry H, Gros C, Barden N (1980) Differential distribution of β-endorphin and enkephalins in rat and bovine brain. Reg Peptides 1:43–52
Dutta AS, Gormley JJ, Hayward CF, Morley JS, Shaw JS, Stacey GJ, Turnbull MT (1977) Enkephalin analogues eliciting analgesia after intravenous injection. Life Sci 21:559–592
Dyhre-Poulsen P (1975) Increased vibration threshold before movements in human subjects. Exp Neurol 47:516–522

Ebersold MJ, Laws ER, Stonnington HH, Stillwell GK (1975) Transcutaneous electrical stimulation for treatment of chronic pain. A preliminary report. Surg Neurol 4:96–98
Eccles JC (1931) Studies on the flexor reflex. III. The central effects produced by an antidromic volley. Proc R Soc Lond [Biol] 107:557–585
Eccles JC (1964) The physiology of synapses. Springer New York Berlin Heidelberg
Eccles JC, Granit R (1929) Crossed extensor reflexes and their interaction. J Physiol (Lond) 67:97–118
Eccles JC, Sherrington CS (1930) Reflex summation in the ipsilateral spinal flexion reflex. J Physiol (Lond) 69:1–28
Eccles JC, Sherrington C (1931a) Studies on the flexor reflex. I. Latent period. Proc R Soc Lond [Biol] 107:511–534
Eccles JC, Sherrington CS (1931b) Studies on the flexor reflex. II. The reflex response evoked by two centripetal volleys. Proc R Soc Lond [Biol] 107:535–556
Eccles JC, Sherrington C (1931c) Studies on the flexor reflex. IV. After-discharge. Proc R Soc Lond [Biol] 107:586–596
Eccles JC, Sherrington C (1931d) Studies on the flexor reflex. V. General conclusions. Proc R Soc Lond [Biol] 107:597–605
Eccles JC, Sherrington C (1931e) Studies on the flexor reflex. VI. Inhibition. Proc R Soc Lond [Biol] 109:91–113
Eccles JC, Eccles RM, Lundberg A (1957) The convergence of monosynaptic excitatory afferents on to many different species of alpha motoneurones. J Physiol (Lond) 137:22–50
Eccles JC, Eccles RM, Lundberg A (1960) Types of neurone in and around the intermediate nucleus of the lumbosacral cord. J Physiol (Lond) 154:89–114
Eccles JC, Kostyuk PG, Schmidt RF (1962) Presynaptic inhibition of the central actions of flexor reflex afferents. J Physiol (Lond) 161:258–281
Eccles JC, Schmidt RF, Willis WD (1963) Depolarization of the central terminals in cutaneous afferent fibers. J Neurophysiol 26:646–661
Eccles RM, Lundberg A (1958) Significance of supraspinal control of reflex actions by impulses in muscle afferents. Experientia 14:197–199
Eccles RM, Lundberg A (1959a) Synaptic actions in motoneurones by afferents which may evoke the flexion reflex. Arch Ital Biol 97:199–221
Eccles RM, Lundberg A (1959b) Supraspinal control of interneurones mediating spinal reflexes. J Physiol (Lond) 147:565–584
Eddy NB, Leimbach D (1953) Synthetic analgesics. II. Dithienylbutenyl- and dithienylbutylamines. J Pharmacol Exp Ther 107:385–393
Eddy NB, Touchberry CF, Lieberman JE (1950) Synthetic analgesics. I. Methadone isomers and derivatives. J Pharmacol Exp Ther 98:121–137
Edwards SB (1975) Autoradiographic studies of the projections of the midbrain reticular formation: descending projections of nucleus cuneiformis. J Comp Neurol 161:341–358
Ehrenpreis S, Balagot RC, Comaty JE, Myles SB (1979) Naloxone reversible analgesia in mice produced by D-phenylalanine and hydrocinnamic acid, inhibitors of carboxypeptidas A. Adv Pain Res Ther 3:479–488
El-Sobky A, Dostrovsky JO, Wall PD (1976) Lack of effect of naloxone on pain perception in humans. Nature 263:783–784
Ekholm J, Eklund G, Skoglund S (1960) On the reflex effects from the knee joint of the cat. Acta Physiol Scand 50:167–174
Elde R, Hökfelt T, Johansson O, Terenius L (1976) Immunohistochemical studies using antibodies to leucine-enkephalin: Initial observations on the nervous system of the rat. Neuroscience 1:349–351
Engberg I, Lundberg A (1969) An electromyographic analysis of muscular activity in the hindlimb of the cat during unrestrained locomotion. Acta Physiol Scand 75:614–630

Engberg I, Ryall RW (1966) The inhibitory action of noradrenaline and other monoamines on spinal neurones. J Physiol (Lond) 185:298–322

Engberg I, Lundberg A, Ryall RW (1965) Reticulospinal inhibition of transmission through interneurones of spinal reflex pathways. Experientia 21:612–613

Engberg I, Lundberg A, Ryall RW (1968a) The effect of reserpine on transmission in the spinal cord. Acta Physiol Scand 72:115–122

Engberg I, Lundberg A, Ryall RW (1968b) Is the tonic decerebrate inhibition of reflex paths mediated by monoaminergic pathways? Acta Physiol Scand 72:123–133

Engberg I, Lundberg A, Ryall RW (1968c) Reticulospinal inhibition of transmission in reflex pathways. J Physiol (Lond) 194:201–223

Engberg I, Lundberg A, Ryall RW (1968d) Reticulospinal inhibition of interneurones. J Physiol (Lond) 194:225–236

Eriksson M, Sjölund B (1976) Acupuncturelike electroanalgesia in TNS-resistant chronic pain. In Zotterman Y (ed) Sensory functions of the skin. Pergamon, New York, pp 575–580

Eriksson MBE, Sjölund BH, Nielzén S (1979) Long term results of peripheral conditioning stimulation as an analgesic measure in chronic pain. Pain 6:335–347

Evans FJ (1974) The placebo response in pain reduction. Adv Neurol 4:289–296

Evans WO (1961) A new technique for the investigation of some analgesic drugs on a reflexive behavior in the rat. Psychopharmacologia 2:318–325

Evarts EV (1971) Feedback and corollary discharge: a merging of the concepts. Neurosci Res Program Bull 9:86–112

Favale E, Loeb C, Manfredi M, Sacco G (1965) Somatic afferent transmission and cortical responsiveness during natural sleep and arousal in the cat. EEG Clin Neurophysiol 18:354–368

Feather BW, Chapman CR, Fisher SB (1972) The effect of a placebo on the perception of painful radiant heat stimuli. Psychsom Med 34:290–294

Fedina L, Lundberg A, Vyklický L (1971) The effect of a noradrenaline liberator (4, alpha-dimethyl-meta-tyramine) on reflex transmission in spinal cats. Acta Physiol Scand 83:495–504

Feldberg W, Smyth DG (1977) C fragment of lipotropin – an endogenous potent analgesic peptide. Br J Pharmacol 60:445–453

Felten DL, Laties AM, Carpenter MB (1974) Monoamine-containing cell bodies in the squirrel monkey brain. Am J Anat 139:153–166

Fennessy MR, Lee JR (1970) Modification of morphine analgesia by drugs affecting adrenergic and tryptaminergic mechanisms. J Pharm Pharmacol 22:930–935

Fetz EE (1968) Pyramidal tract effect on interneurons in the cat lumbar dorsal horn. J Neurophysiol 31:69–80

Fibiger HC, Mertz PH, Campbell BA (1972) The effect of para-chlorophenylalanine on aversion thresholds and reactivity to foot shock. Physiol Behav 8:259–263

Fields HL, Adams JE (1974) Pain after cortical injury relieved by electrical stimulation of the internal capsule. Brain 97:169–178

Fields HL, Anderson SD (1978) Evidence that raphe-spinal neurons mediate opiate and midbrain stimulation-produced analgesias. Pain 5:333–349

Fields HL, Basbaum AI (1978) Brainstem control of spinal pain-transmission neurons. Annu Rev Physiol 40:217–248

Fields HL, Adams JE, Hosobuchi Y (1974) Peripheral nerve and cutaneous electrohypalgesia. Adv Neurol 4:749–754

Fields HL, Wagner GM, Anderson SD (1975) Some properties of spinal neurons projecting to the medial brain-stem reticular formation. Exp Neurol 47:118–134

Fields HL, Basbaum AI, Clanton CH, Anderson SD (1977a) Nucleus raphe magnus inhibition of spinal cord dorsal horn neurons. Brain Res 126:441–453

Fields HL, Clanton CH, Anderson SD (1977b) Somatosensory properties of spinoreticular neurons in the cat. Brain Res 120:49–66

Fields HL, Emson PC, Leigh BK, Gilbert RFT, Iversen LL (1980) Multiple opiate receptor sites on primary afferent fibers. Nature 284:351–353

Finer B (1974) Clinical use of hypnosis in pain management. Adv Neurol 4:573–579

Finer BL, Nylén BO (1961) Cardiac arrest in the treatment of burns, and report on hypnosis as a substitute for anesthesia. Plast Reconstr Surg 27:49–55

Flodmark S, Wramner T (1945) The analgetic action of morphine, eserine and prostigmine studied by a modified Hardy-Wolff-Goodell method. Acta Physiol Scand 9:88–96

Foerster O, Gagel O (1932) Die Vorderseitenstrangdurchschneidung beim Menschen. Eine klinisch-patho-physiologisch-anatomische Studie. Z Ges Neurol Psychiat 138:1–92

Forbes A, Cobb S, Cattell H (1923) Electrical studies in mammalian reflexes. III. Immediate changes in the flexion reflex after spinal transection. Am J Physiol 65:30–44

Forbes A, Cattell McK (1924) Electrical studies in mammalian reflexes. IV. The crossed extension reflex. Am J Physiol 70:140–173

Forbes A, Gregg A (1915) Electrical studies in mammalian reflex. I. The flexion reflex. Am J Physiol 37:118–176

Foreman RD, Weber RN (1980) Responses from neurons of the primate spinothalamic tract to electrical stimulation of afferents from the cardiopulmonary region and somatic structures. Brain Res 186:463–468

Foreman RD, Applebaum AE, Beall JE, Trevino DL, Willis WD (1975) Responses of primate spinothalamic tract neurons to electrical stimulation of hindlimb peripheral nerves. J Neurophysiol 38:132–145

Foreman RD, Beall JE, Applebaum AE, Coulter JD, Willis WD (1976) Effects of dorsal column stimulation on primate spinothalamic tract neurons. J Neurophysiol 39:534–546

Foreman RD, Schmidt RF, Willis WD (1977) Convergence of muscle and cutaneous input onto primate spinothalamic tract neurons. Brain Res 124:555–560

Foreman RD, Kenshalo DR, Jr, Schmidt RF, Willis WD (1979a) Field potentials and excitation of primate spinothalamic neurones in response to volleys in muscle afferents. J Physiol (Lond) 286:197–213

Foreman RD, Schmidt RF, Willis WD (1979b) Effects of mechanical and chemical stimulation of fine muscle afferents upon primate spinothalamic tract cells. J Physiol (Lond) 286:215–231

Foreman RD, Hancock MB, Willis WD (1981) Responses of spinothalamic tract cells in the thoracic spinal cord of the monkey to cutaneous and visceral inputs. Pain 11:149–162

Fox EJ, Melzack R (1976) Transcutaneous electrical stimulation and acupuncture: comparison of treatment for low-back pain. Pain 2:141–148

Fox JL (1974) Dorsal column stimulation for relief of intractable pain: problems encountered with neuropacemakers. Surg Neurol 2:59–64

Franz DN, Iggo A (1968) Dorsal root potentials and ventral root reflexes evoked by nonmyelinated fibers. Science 162:1140–1142

Frederickson RCA, Burgis V, Edwards JD (1977) Hyperalgesia induced by naloxone follows diurnal rhythm in responsivity to painful stimuli. Science 198:756–758

Fu TC, Halenda SP, Dewey WL (1980) The effect of hypophysectomy on acupuncture analgesia. Brain Res 202:33–39

Fukushima K, Kato M (1975) Spinal interneurons responding to group II muscle afferent fibers in the cat. Brain Res 90:307–312

Fulton JF (1926) Muscular contraction and the reflex control of movement. Williams & Wilkins, Baltimore, p 644

Fung DTH, Hwang JC, Chan SHH, Chin YC (1975) Electro-acupuncture suppression of jaw depression reflex elicited by dentalgia in rabbits. Exp Neurol 47:367–369

Fung DTH, Hwang JC, Chan SHH, Chan YC (1976) Jaw-opening reflex as pain response in rabbit: preliminary report. J Dent Res 55:917

Fung DT, Hwang JC, Chung SH (1978) Correlative study of pain perception and masticatory muscle reflexes in man. Oral Surg 45:44–50
Galambos R (1956) Suppression of auditory nerve activity by stimulation of efferent fibers to cochlea. J Neurophysiol 19:424–437
Gallager DW, Pert A (1978) Afferents to brain stem nuclei (brain stem raphe, nucleus reticularis pontis caudalis and nucleus gigantocellularis) in the rat as demonstrated by microiontophoretically applied horseradish peroxidase. Brain Res 144:257–275
Galeano C, Leung CY (1978) Has acupuncture an analgesic effect in the rabbit? Pain 4:265–271
Galeano C, Leung CY, Robitaille R, Roy-Chabot T (1979) Acupuncture analgesia in rabbits. Pain 6:71–81
Gammon GD, Starr I (1941) Studies on the relief of pain by counterirritation. J Clin Invest 20:13–20
Garau L, Mulas ML, Pepeu G (1975) The influence of raphe lesions on the effect of morphine on nociception and cortical ACh output. Neuropharmacol 14:259–263
Garcia Leme J, Rocha e Silva M (1961) Analgesic action of chlorpromazine and reserpine in relation to that of morphine. J Pharm Pharmacol 13:734–742
Gardner E (1950) Reflex muscular responses to stimulation of articular nerves in the cat. Am J Physiol 161:133–141
Gaw AC, Chang LW, Shaw LC (1975) Efficacy of acupuncture on osteoarthritic pain. N Engl J Med 293:375–378
Gebhart GF, Toleikis JR (1978) An evaluation of stimulation-produced analgesia in the cat. Exp Neurol 62:570–579
Genovese E, Zonta N, Mantegazza P (1973) Decreased antinociceptive activity of morphine in rats pretreated intraventricularly with 5,6-dihydroxytryptamine, a long-lasting selective depletor of brain serotonin. Psychopharmacologia 32:359–364
Gerhart KD, Wilcox TK, Chung JM, Willis WD (1981a) Inhibition of nociceptive and nonnociceptive responses of primate spinothalamic cells by stimulation in medial brain stem. J Neurophysiol 45:121–136
Gerhart KD, Yezierski RP, Giesler GJ, Willis WD (1981b) Inhibitory receptive fields of primate spinothalamic tract cells. J Neurophysiol 46:1309–1325
Gerhart KD, Yezierski RP, Wilcox TK, Grossman AE, Willis WD (1981c) Inhibition of primate spinothalamic tract neurons by stimulation in ipsilateral or contralateral ventral posterior lateral (VPL_c) thalamic nucleus. Brain Res 229:514–519
Ghez C, Lenzi GL (1971) Modulation of sensory transmission in cat lemniscal system during voluntary movement. Pfluegers Arch 323:273–278
Ghez C, Pisa M (1972) Inhibition of afferent transmission in cuneate nucleus during voluntary movement in the cat. Brain Res 40:145–151
Ghia JN, Mao W, Toomey TC, Gregg JM (1976) Acupuncture and chronic pain mechanisms. Pain 2:285–299
Gibson JJ (1966) The senses considered as perceptual systems. Houghton Mifflin Co, Boston
Giesler GJ, Liebeskind JC (1976) Inhibition of visceral pain by electrical stimulation of the periaqueductal gray matter. Pain 2:43–48
Giesler GJ, Menétrey D, Guilbaud G, Besson JM (1976) Lumbar cord neurons at the origin of the spinothalamic tract in the rat. Brain Res 118:320–324
Giesler GJ, Gerhart KD, Yezierski RP, Wilcox TK, Willis WD (1981a) Postsynaptic inhibition of primate spinothalamic neurons by stimulation in nucleus raphe magnus. Brain Res 204:184–188
Giesler GJ, Yezierski RP, Gerhart KD, Willis WD (1981b) Spinothalamic tract neurons that project to medial and/or lateral thalamic nuclei: evidence for a physiologically novel population of spinal cord neurons. J Neurophysiol 46:1285–1308

Gilbert PE, Martin WR (1976) The effects of morphine- and nalorphine-like drugs in the nondependent, morphine-dependent and cyclazocine-dependent chronic spinal dog. J Pharmacol Exp Ther 198:66 – 82
Gillan MGC, Kosterlitz HW, Paterson SJ (1979) Comparison of the binding characteristics of tritiated opiates and opioid peptides. Br J Pharmacol 66:86 – 87P
Glazer EJ, Basbaum AI (1981) Immunohistochemical localization of leucine-enkephalin in the spinal cord of the cat: enkephalin-containing marginal neurons and pain modulation. J Comp Neurol 196:377 – 389
Gobel S, Falls WM, Bennett GJ, Abdelmoumene M, Hayashi H, Humphrey E (1980) An EM analysis of the synaptic connections of horseradish peroxidase-filled stalked cells and islet cells in the substantia gelatinosa of adult cat spinal cord. J Comp Neurol 194:781 – 807
Gol A (1967) Relief of pain by electrical stimulation of the septal area. J Neurol Sci 5:115 – 120
Goldberger SM, Tursky B (1976) Modulation of shock-elicited pain by acupuncture and suggestion. Pain 2:417 – 429
Goldstein A (1976) Opioid peptides (endorphins) in pituitary and brain. Science 193:1081 – 1086
Goldstein A, Hilgard ER (1975) Failure of the opiate antagonist naloxone to modify hypnotic analgesia. Proc Natl Acad Sci USA 72:2041 – 2043
Goldstein A, Pryor GT, Otis LS, Larsen F (1976) On the role of endogenous opioid peptides: Failure of naloxone to influence shock escape threshold in the rat. Life Sci 18:599 – 604
Goldstein A, Tachibana S, Lowney LI, Hunkapiller M, Hood L (1979) Dynorphin (1 – 13), an extraordinarily potent opioid peptide. Proc Natl Acad Sci USA 76:6666 – 6670
Goode GE, Humbertson AO, Martin GF (1980) Projections from the brain stem reticular formation to laminae I and II of the spinal cord. Studies using light and electron microscopic techniques in the North American opossum. Brain Res 189:327 – 342
Goodlet I, Sugrue MF (1974) effects of acutely administered analgesic drugs on rat brain serotonin turnover. Eur J Pharmacol 29:241 – 248
Goodman SJ, Holcombe V (1976) Selective and prolonged analgesia in monkey resulting from brain stimulation. Adv Pain Res Ther 1:495 – 502
Gordon G, Jukes MGM (1964) Descending influences on the exteroceptive organization of the cat's gracile nucleus. J Physiol (Lond) 173:291 – 319
Gordon G, Miller R (1969) Identification of cortical cells projecting to the dorsal column nuclei of the cat. Q J Exp Physiol 54:85 – 98
Görlitz BD, Frey HH (1972) Central monoamines and antinociceptive drug action. Eur J Pharmacol 20:171 – 180
Gracely RH, McGrath P, Dubner R (1978a) Ratio scales of sensory and affective verbal pain descriptors. Pain 5:5 – 18
Gracely RH, McGrath P, Dubner R (1978b) Validity and sensitivity of ratio scales of sensory and affective verbal pain descriptors: manipulation of affect by diazepam. Pain 5:19 – 29
Gracely RH, Dubner R, McGrath PA (1979) Narcotic analgesia: fentanyl reduces the intensity but not the unpleasantness of painful tooth pulp sensations. Science 203:1261 – 1263
Gray BG, Dostrovsky JO (1981) Inhibition of cat dorsal horn neurons by noxious cutaneous stimulation of anatomically remote areas of the body surface. Pain [Suppl] 1:S183
Green AF, Young PA, Godfrey EI (1951) A comparison of heat and pressure analgesiometric methods in rats. Br J Pharmacol 6:572 – 585
Gregor M, Zimmermann M (1972) Characteristics of spinal neurones responding to cutaneous myelinated and unmyelinated fibres. J Physiol (Lond) 221:555 – 576
Grevert P, Goldstein A (1977a) Effects of naloxone on experimentally induced ischemic pain and on mood in human subjects. Proc Natl Acad Sci USA 74:1291 – 1294

Grevert P, Goldstein A (1977b) Some effects of naloxone on behavior in the mouse, Psychopharmacology 53:111–113

Grevert P, Goldstein A (1978) Endorphins: Naloxone fails to alter experimental pain or mood in humans. Science 199:1093–1095

Griersmith BT, Duggan AW (1980) Prolonged depression of spinal transmission of nociceptive information by 5-HT administered in the substantia gelatinosa: antagonism by methysergide. Brain Res 187:231–236

Griersmith BT, Duggan AW, North RA (1981) Methysergide and supraspinal inhibition of the spinal transmission of nociceptive information in the anesthetized cat. Brain Res 204:147–158

Grigg P, Harrigan EP, Fogarty KE (1978) Segmental reflexes mediated by joint afferent neurons in cat knee. J Neurophysiol 41:9–14

Grillner S (1970) Is the tonic stretch reflex dependent upon group II excitation? Acta Physiol Scand 78:431–432

Grillner S, Hongo T, Lundberg A (1967) The effect of DOPA on the spinal cord. 7. Reflex activation of static γ-motoneurones from the flexor reflex afferents. Acta Physiol Scand 70:403–411

Gritz ER, Shiffman SM, Jarvik ME, Schlesinger J, Charuvastra VC (1976) Naltrexone: Physiological and psychological effects of single doses. Clin Pharmacol Ther 19:773–776

Grossman W, Jurna I, Nell T, Theres C (1973) The dependence of the anti-nociceptive effect of morphine and other analgesic agents on spinal motor activity after central monoamine depletion. Eur J Pharmacol 24:67–77

Grumbach L (1966) The prediction of analgesic activity in man by animal testing. In: Knighton RS, Dumke PR (eds) Pain. Little Brown, Boston, pp 163–182

Guilbaud G, Benelli G, Besson JM (1977a) Responses of thoracic dorsal horn interneurons to cutaneous stimulation and to the administration of algogenic substances into the mesenteric artery in the spinal cat. Brain Res 124:437–448

Guilbaud G, Besson JM, Oliveras JL, Liebeskind JC (1973) Suppression by LSD of the inhibitory effect exerted by dorsal raphe stimulation on certain spinal cord interneurons in the cat. Brain Res 61:417–422

Guilbaud G, Oliveras JL, Giesler G, Besson JM (1977b) Effects induced by stimulation of the centralis inferior nucleus of the raphe on dorsal horn interneurons in cat's spinal cord. Brain Res 126:355–360

Guillemin R, Ling N, Burgus R, Bloom F, Segal D (1977a) Characterization of the endorphins, novel hypothalamic and neurohypophysial peptides with opiate-like activity: evidences that they induce profound behavioral changes. Psychoneuroendocrinology 2:59–62

Guillemin R, Vargo T, Rossier J, Minick S, Ling N, Rivier C, Vale W, Bloom F (1977b) β-Endorphin and adrenocorticotropin are secreted concomitantly by the pituitary gland. Science 197:1367–1369

Gybels JM (1979) Electrical stimulation of the central gray for pain relief in humans: A critical review. Adv Pain Res Therapy 3:499–509

Ha H, Wu RS, Contreras RA, Tan EC (1978) Measurement of pain threshold by electrical stimulation of tooth pulp afferents in the monkey. Exp Neurol 61:260–269

Ha H, Tan EC, Fukunaga H, Aochi O (1981) Naloxone reversal of acupuncture analgesia in the monkey. Exp Neurol 73:298–303

Haber LH, Martin RF, Chatt AB, Willis WD (1978) Effects of stimulation in nucleus reticularis gigantocellularis on the activity of spinothalamic tract neurons in the monkey. Brain Res 153:163–168

Haber LH, Martin RF, Chung JM, Willis WD (1980) Inhibition and excitation of primate spinothalamic tract neurons by stimulation in region of nucleus reticularis gigantocellularis. J Neurophysiol 43:1578–1593

Haffner F (1929) Experimentelle Prüfung schmerzstillender Mittel. Deutsch Med Wochenschr 55:731 – 733
Hagbarth KE (1952) Excitatory and inhibitory skin areas for flexor and extensor motoneurones. Acta Physiol Scand [Suppl 94] 26:1 – 58
Hagbarth KE (1960) Centrifugal mechanisms of sensory control. Erg Biol 22:47 – 66
Hagbarth KE, Kerr DIB (1954) Central influences on spinal afferent conduction. J Neurophysiol 17:295 – 307
Hagbarth KE, Naess K (1950) Reflex effects of tetanic stimulation of different afferent fibre-systems in the hind limb of the cat. Acta Physiol Scand 21:336 – 361
Haigler HJ, Aghajanian GK (1974) Peripheral serotonin antagonists: failure to antagonize serotonin in brain areas receiving a prominent serotonergic input. J Neural Transm 35:257 – 273
Halpern M (1968) Effects of midbrain central gray matter lesions on escape-avoidance behavior in rats. Physiol Behav 3:171 – 178
Hammond DL, Levy RA, Proudfit HK (1980) Hypoalgesia induced by microinjection of a norepinephrine antagonist in the raphe magnus: reversal by intrathecal administration of a serotonin antagonist. Brain Res 201:475 – 479
Han CS, Chou PH, Lu CC, Lu LH, Yang TH, Jen MF (1979) The role of central 5-hydroxytryptamine in acupuncture analgesia. Sci Sin 22:91 – 104
Han JS, Tang J, Ren MF, Zhou ZF, Fan SG, Qiu XC (1980) Central neurotransmitters and acupuncture analgesia. Am J Chin Med 8:331 – 348
Hancock MB, Fougerousse CL (1976) Spinal projections from the nucleus locus coeruleus and nucleus subcoeruleus in the cat and monkey as demonstrated by the retrograde transport of horse radish peroxidase. Brain Res Bull 1:229 – 234
Hancock MB, Foreman RD, Willis WD (1975) Convergence of visceral and cutaneous input onto spinothalamic tract cells in the thoracic spinal cord of the cat. Exp Neurol 47:240 – 248
Handwerker HO, Iggo A, Zimmermann M (1975) Segmental and supraspinal actions on dorsal horn neurons responding to noxious and non-noxious skin stimuli. Pain 1:147 – 165
Hardy JD (1953) Thresholds of pain and reflex contraction as related to noxious stimulation. J Appl Physiol 5:725 – 739
Hardy JD (1962) The pain threshold and the nature of pain sensation. In Keele CA, Smith R: The assessment of pain in man and animals. Livingstone, London, pp 170 – 212
Hardy JD, Wolff HG, Goodell H (1940) Studies on pain. A new method for measuring pain threshold: observations on spatial summation of pain. J Clin Invest 19:649 – 657
Hardy JD, Wolff HG, Goodell H (1952) Pain sensations and reactions. Williams & Wilkins, New York, (reprinted by Hafner Publishing Co, 1967)
Harrison IB, Bigelow NH (1943) Quantitative studies of visceral pain produced by the contraction of ischemic muscle. Proc Assoc Res Nerve Ment Dis 23:154 – 165
Harvey JA, Lints CE (1971) Lesions in the medial forebrain bundle: relationship between pain sensitivity and telencephalic content of serotonin. J Comp Physiol Psychol 74:28 – 36
Haubrich DR, Blake DE (1973) Modification of serotonin metabolism in rat brain after acute or chronic administration of morphine. Biochem Pharmacol 22:2753 – 2759
Hayashi G, Takemori AE (1971) The type of analgesic-receptor interaction involved in certain analgesic assays. Eur J Pharmacol 16:63 – 66
Hayes NL, Rustioni A (1981) Descending projections from brainstem and sensorimotor cortex to spinal enlargements in the cat. Exp Brain Res 41:89 – 107
Hayes RL, Bennett GJ, Mayer DJ (1975) Acupuncture, pain, and signal detection theory Science 189:65 – 66

Hayes RL, Newlon PG, Rosecrans JA, Mayer DJ (1977a) Reduction of stimulation-produced analgesia by lysergic acid diethylamide, a depressor of serotonergic neural activity. Brain Res 122:367 – 372
Hayes RL, Price DD, Dubner R (1977b) Naloxone antagonism as evidence for narcotic mechanisms. Science 196:600
Hayes RL, Bennett GJ, Newlon PG, Mayer DJ (1978) Behavioral and physiological studies of non-narcotic analgesia in the rat elicited by certain environmental stimuli. Brain Res 155:69 – 90
Hayes RL, Price DD, Bennett GL, Wilcox GL, Mayer DJ (1979a) Differential effects of spinal cord lesions on narcotic and non-narcotic suppression of nociceptive reflexes: Further evidence for the physiological multiplicity of pain modulation. Brain Res 155:91 – 101
Hayes RL, Price DD, Dubner R (1979b) Behavioral and physiological studies of sensory coding and modulation of trigeminal nociceptive input. Adv Pain Res Ther 3:219 – 243
Hayes RL, Price DD, Ruda M, Dubner R (1979c) Suppression of nociceptive responses in the primate by electrical stimulation of the brain or morphine administration: behavioral and electrophysiological comparisons. Brain Res 167:417 – 421
Hayes RL, Dubner R, Hoffman DS (1981) Neuronal activity in medullary dorsal horn of awake monkeys trained in a thermal discrimination task. II. Behavioral modulation of responses to thermal and mechanical stimuli. J Neurophysiol 46:428 – 443
Head H, Holmes G (1911) Sensory disturbances from cerebral lesions. Brain 34:102 – 254
Headley PM, Duggan AW, Griersmith BT (1978) Selective reduction by noradrenaline and 5-hydroxytryptamine of nociceptive responses of cat dorsal horn neurones. Brain Res 145:185 – 189
Heavner JE, DeJong RH (1973) Spinal cord neuron response to natural stimuli. A microelectrode study. Exp Neurol 39:293 – 306
Hendershot LC, J Forsaith (1959) Antagonism of the frequency of phenylquinone-induced writhing in the mouse by weak analgesics and non-analgesics. J Pharmacol Exp Ther 125:237 – 240
Hentall ID, Fields HL (1979) Segmental and descending influences on intraspinal thresholds of single C-fibres. J Neurophysiol 42:1527 – 1537
Hernández-Peón R (1955) Central mechanisms controlling conduction along central sensory pathways. Acta Neurol Lat Am 1:256 – 264
Hernández-Peón R, Scherrer H, Jouvet M (1956a) Modification of electric activity in cochlear nucleus during "attention" in unanaesthetized cats. Science 123:331 – 332
Hernández-Peón R, Scherrer H, Velasco M (1956b) Central influences on afferent conduction in the somatic and visual pathways. Acta Neurol Lat Am 2:8 – 22
Hernández-Peón R, Jouvet M, Scherrer H (1957) Auditory potentials at cochlear nucleus during acoustic habituation. Acta Neurol Lat Am 3:144 – 156
Herz A, Albus K, Metys J, Schubert P, Teschemacher H (1970) On the central sites for the antinociceptive action of morphine and fentanyl. Neuropharmacol 9:539 – 551
Herz A, Teschemacher H (1971) Activities and sites of antinociceptive action of morphine-like analgesics and kinetics of distribution following intravenous, intracerebral and intraventricular application. Adv Drug Res 6:79 – 119
Hiedl P, Struppler A, Gessler M (1979) Local analgesia by percutaneous electrical stimulation of sensory nerves. Pain 7:129 – 134
Hilgard ER (1975) The alleviation of pain by hypnosis. Pain 1:213 – 231
Hilgard ER, Hilgard JR (1975) Hypnosis in the relief of pain. W Kaufmann, Los Altos California
Hill RG, Ayliffe SJ (1981) The antinociceptive effect of vaginal stimulation in the rat is reduced by naloxone. Pharmacol Biochem Behav 14:631 – 632
Hiller JM, Pearson J, Simon EJ (1973) Distribution of stereospecific binding of the potent narcotic analgesic etorphine in the human brain: predominance in the limbic system. Res Comm Chem Pathol Pharmacol 6:1052 – 1062

Hillman P, Wall PD (1969) Inhibitory and excitatory factors influencing the receptive fields of lamina 5 spinal cord cells. Exp Brain Res 9:284 – 306

Ho IK, Brase DA, Loh HH, Way EL (1975) Influence of L-tryptophan on morphine analgesia, tolerance, and physical dependence. J Pharmacol Exp Ther 193:35 – 43

Hodge CJ, Apkarian AV, Stevens R, Vogelsang G, Wisnicki HJ (1981) Locus coeruleus modulation of dorsal horn unit responses to cutaneous stimulation. Brain Res 204:514 – 420

Hoff C (1980) Immoral and moral uses of animals. N Engl J Med 302:115 – 118

Hökfelt T, Kellerth JO, Nilsson G, Pernow B (1975) Experimental immunohistochemical studies on the localization and distribution of substance P in cat primary sensory neurons. Brain Res 100:235 – 252

Hökfelt T, Elde R, Johansson O, Luft R, Nilsson G, Arimura A (1976) Immunohistochemical evidence for separate populations of somatostatin-containing and substance P-containing primary afferent neurons in the rat. Neuroscience 1:131 – 136

Hökfelt T, Elde R, Johansson O, Terenius L, Stein L (1977a) The distribution of enkephalin-immunoreactive cell bodies in the rat central nervous system. Neurosci Lett 5:25 – 31

Hökfelt T, Ljungdahl A, Terenius L, Elde R, Nilsson G (1977b) Immunohistochemical analysis of peptide pathways possibly related to pain and analgesia: Enkephalin and substance P. Proc Natl Acad Sci USA 74:3081 – 3085

Hökfelt T, Ljungdahl A, Steinbusch H, Verhofstad A, Nilsson G, Brodin E, Pernow B, Goldstein M (1978) Immunohistochemical evidence of substance P-like immunoreactivity in some 5-hydroxytryptamine-containing neurons in the rat central nervous system. Neuroscience 3:517 – 538

Hökfelt T, Terenius T, Kuypers HGJM, Dann O (1979) Evidence for enkephalin immunoreactive neurons in the medulla oblongata projecting to the spinal cord. Neuroscience Lett 14:55 – 60

Holmgren E (1975) Increase of pain threshold as a function of conditioning electrical stimulation. Am J Chin Med 3:133 – 142

Holmqvist B (1961) Crossed spinal reflex actions evoked by volleys in somatic afferents. Acta Physiol Scand [Suppl 181] 52:1 – 67

Holmqvist B, Lundberg A (1959) On the organization of the supraspinal inhibitory control of interneurones of various spinal reflex arcs. Arch Ital Biol 97:340 – 356

Holmqvist B, Lundberg A (1961) Differential supraspinal control of synaptic actions evoked by volleys in the flexion reflex afferents in alpha motoneurones. Acta Physiol Scand [Suppl. 186] 54:1 – 51

Holmqvist B, Lundberg A, Oscarsson O (1960) Supraspinal inhibitory control of transmission to three ascending spinal pathways influenced by the flexion reflex afferents. Arch Ital Biol 98:60 – 80

Holst E von (1954) Relations between the central nervous system and the peripheral organs. Br J Animal Behav 2:89 – 94

Hong JS, Yang HYT, Fratta W, Costa E (1977) Determination of methionine enkephalin in discrete regions of rat brain. Brain Res 134:383 – 386

Hong SK, Kniffki KD, Mense S, Schmidt RF, Wendisch M (1979) Descending influences on the responses of spinocervical tract neurones to chemical stimulation of fine muscle afferents. J Physiol (Lond) 290:129 – 140

Hongo T, Jankowska E, Lundberg A (1966) Convergence of excitatory and inhibitory action on interneurones in the lumbosacral cord. Exp Brain Res 1:338 – 358

Hosobuchi Y (1978) Tryptophan reversal of tolerance to analgesia induced by central grey stimulation. Lancet 11:47

Hosobuchi Y, Adams JE, Rutkin B (1973) Chronic thalamic stimulation for the control of facial anesthesia dolorosa. Arch Neurol 29:158 – 161

Hosobuchi Y, Adams JE, Fields HL (1974) Chronic thalamic and internal capsular stimulation for the control of facial anesthesia dolorosa and dysesthesia of thalamic syndrome. Adv Neurol 4:783 – 787

Hosobuchi Y, Adams JE, Linchitz R (1977a) Pain relief by electrical stimulation of the central gray matter in humans and its reversal by naloxone. Science 197:183 – 186
Hosobuchi Y, Lamb S, Bascom D (1980) Tryptophan loading may reverse tolerance to opiate analgesics in humans: a preliminary report. Pain 9:161 – 169
Hosobuchi Y, Meglio M, Adams JE, Li CH (1977b) β-Endorphin: development of tolerance and its reversal by 5-hydroxytryptophan in cats. Proc Natl Acad Sci USA 74:4017 – 4019
Hosobuchi Y, Rossier J, Bloom FE, Guillemin R (1979) Stimulation of human periaqueductal gray for pain relief increases immunoreactive β-endorphin in ventricular fluid. Science 203:279 – 281
Houde RW (1966) On assaying analgesics in man. In Knighton RS, Dumke PR (eds): Pain. Little, Brown & Co, Boston, pp 183 – 196
Houde RW, Wallenstein SL, Rogers A (1960) Clinical pharmacology of analgesics. 1. A method of assaying analgesic effect. Clin Pharmacol Ther 1:163 – 174
Hubbard JE, Di Carlo V (1974) Fluorescence histochemistry of monoamine-containing cell bodies in the brain stem of the squirrel monkey (Saimiri sciureus). III. Serotonin-containing groups. J Comp Neurol 153:385 – 398
Hughes J (1975) Isolation of an endogenous compound from the brain with pharmacological properties similar to morphine. Brain Res 88:295 – 308
Hughes J, Smith TW, Kosterlitz HW, Fothergill LA, Morgan BA, Morris HR (1975) Identification of two related pentapeptides from the brain with potent opiate agonist activity. Nature 258:557 – 579
Hughes J, Kosterlitz HW, Smith TW (1977) The distribution of methionine-enkephalin and leucine-enkephalin in the brain and peripheral tissues. Br J Pharmacol 61:639 – 647
Hunt CC (1954) Relation of function to diameter in afferent fibers of muscle nerves. J Gen Physiol 38:117 – 131
Hunt CC, Perl ER (1960) Spinal reflex mechanisms concerned with skeletal muscle. Physiol Rev 40:538 – 579
Hunt SP, Kelly JS, Emson PC (1980) The electron microscopic localization of methionine-enkephalin within the superficial layers (I and II) of the spinal cord. Neuroscience 5:1871 – 1890
Huskisson EC (1974) Measurement of pain. Lancet 2:1127 – 1131
Iggo A (1979) Experimental study of pain in animals – Ethical aspects. Adv Pain Ther 3:773 – 778
Ignelzi RJ, Nyquist JK (1976) Direct effect of electrical stimulation on peripheral nerve evoked activity: implications in pain relief. J Neurosurg 45:159 – 165
Ireson JD (1970) A comparison of the antinociceptive actions of cholinomimetic and morphine-like drugs. Br J Pharmacol 40:92 – 101
Irwin S, Houde RW, Bennett DR, Hendershot LC, Steevers MH (1951) The effects of morphine, methadone and meperidine on some reflex responses of spinal animals to nociceptive stimulation. J Pharmacol Exp Ther 101:132 – 143
Jabbur SJ, Towe AL (1961) Cortical excitation of neurons in dorsal column nuclei of cat, including an analysis of pathways. J Neurophysiol 24:499 – 509
Jacob JJ, Tremblay EC, Colombel MC (1974) Facilitation de réactions nociceptives par la naloxone chez la souris et chez le rat. Psychopharmacologia 37:217 – 223
Jacquet YF, Lajtha A (1973) Morphine action at central nervous system sites in rat: analgesia or hyperalgesia depending on site and dose. Science 182:490 – 492
Jacquet YF, Lajtha A (1974) Paradoxical effects after microinjection of morphine in the periaqueductal gray matter in the rat. Science 185:1055 – 1057
Jacquet YF, Lajtha A (1976) The periaqueductal grey: sites of morphine analgesia and tolerance as shown by 2-way cross tolerance between systemic and intracerebral injections. Brain Res 103:501 – 513

Janko M, Trontelj JV (1980) Transcutaneous electrical nerve stimulation: A microneurographic and perceptual study. Pain 9:219 – 230
Jankowska E, Lindström S (1971) Morphological identification of Renshaw cells. Acta Physiol Scand 81:428 – 430
Jankowska E, Lindström S (1972) Morphology of interneurones mediating Ia reciprocal inhibition of motoneurones in the spinal cord of the cat. J Physiol (Lond) 226:805 – 823
Jankowska E, Roberts WJ (1972a) An electrophysiological demonstration of the axonal projections of single spinal interneurones in the cat. J Physiol (Lond) 222:597 – 622
Jankowska E, Roberts WJ (1972b) Synaptic actions of single interneurones mediating reciprocal Ia inhibition of motoneurones. J Physiol (Lond) 222:623 – 642
Jankowska E, Lund S, Lundberg A (1966) The effect of DOPA on the spinal cord. 4. Depolarization evoked in the central terminals of contralateral Ia afferent terminals by volleys in the flexor reflex afferents. Acta Physiol Scand 68:337 – 341
Jankowska E, Jukes MGM, Lund S, Lundberg A (1967a) The effect of DOPA on the spinal cord. 5. Reciprocal organization of pathways transmitting excitatory action to alpha motoneurones of flexors and extensors. Acta Physiol Scand 70:369 – 388
Jankowska E, Jukes MGM, Lund S, Lundberg A (1967b) The effect of DOPA on the spinal cord. 6. Half-centre organization of interneurones transmitting effects from the flexor reflex afferents. Acta Physiol Scand 70:389 – 402
Jankowska E, Lund S, Lundberg A, Pompeiano O (1968) Inhibitory effects evoked through ventral reticulospinal pathways. Arch Ital Biol 106:124 – 140
Jarvik ME, Wolff BB (1962) Differences between deep pain responses to hypertonic and hypotonic saline solutions. J Appl Physiol 17:841 – 843
Jessell TM, Iversen LL (1977) Opiate analgesics inhibit substance P release from rat trigeminal nucleus. Nature 268:549 – 551
Job C (1953) Über autogene Inhibition und Reflexumkehr bei spinalisierten und decerebrierten Katzen. Pfluegers Arch 256:406 – 418
Johannesson T, Woods LA (1964) Analgesic action and brain and plasma levels of morphine and codeine in morphine tolerant, codeine tolerant and non-tolerant rats. Acta Pharmacol Toxicol (Copenh) 21:381 – 396
Johns D, DeLanerolle N, LaMotte C (1981) Localization of 5-HT in monkey spinal cord using light and EM immunohistochemistry. Anat Rec 199:129A
Jones B (1979) Signal detection theory and pain research. Pain 7:305 – 312
Johnson KO (1974) Reconstruction of population response to a vibratory stimulus in quickly adapting mechanoreceptive afferent fiber population innervating glabrous skin of the monkey. J Neurophysiol 37:48 – 72
Johnson KO, Darian-Smith I, LaMotte C (1973) Peripheral neural determinants of temperature discrimination in man: A correlative study of responses to cooling skin. J Neurophysiol 36:347 – 370
Jordan LM, Willis WD (1972) Effects of para-methoxyphenylethylamine on reflexes and motoneurons in the cat lumbar spinal cord. J Pharm Exp Ther 181:53 – 64
Jordan LM, Willis WD (1973) Effect of para-methoxyphenylethylamine on polysynaptic pathways and on interneurons in the cat lumbar spinal cord. J Pharm Exp Ther 185:572 – 582
Jordan LM, McCrea DA (1976) Analysis of the effects of *p*-methoxyphenylethylamine on spinal cord neurones. Br J Pharmacol 57:191 – 199
Jordan LM, Kenshalo DR, Jr, Martin RF, Haber LH, Willis WD (1978) Depression of primate spinothalamic tract neurons by iontophoretic application of 5-hydroxytryptamine. Pain 5:135 – 142
Jordan LM, Kenshalo DR, Jr, Martin RF, Haber LH, Willis WD (1979) Two populations of spinothalamic tract neurons with opposite responses to 5-hydroxytryptamine. Brain Res 164:342 – 346

Joyce CRB, Zutshi DW, Hrubes V, Mason RM (1975) Comparison of fixed interval and visual analogue scales for rating chronic pain. Eur J Clin Pharmacol 8:415 – 420
Jurna I (1980) Effect of stimulation in the periaqueductal grey matter on activity in ascending axons of the rat spinal cord: selective inhibition of activity evoked by afferent Aδ and C fibre stimulation and failure of naloxone to reduce inhibition. Brain Res 196:33 – 42
Kaada B (1974) Mechanisms of acupuncture analgesia. T norske Laegeforen 94:422 – 431
Kaada B (1976) Neurophysiology and acupuncture: a review. Adv Pain Res Therap 1:733 – 741
Kane K, Taub A (1975) A history of local electrical analgesia. Pain 1:125 – 138
Kao FF (1973) China, Chinese medicine, and the Chinese medical system. Am J Chin Med 1:1 – 59
Kayan S, Woods LA, Mitchell CL (1969) Experience as a factor in the development of tolerance to the analgesic effect of morphine. Eur J Pharmacol 6:333 – 339
Kellaway P (1946) The part played by electric fish in the early history of bioelectricity and electrotherapy. Bull Hist Med 20:112 – 137
Kelly DD, Glusman M (1968) Aversive thresholds following midbrain lesions. J Comp Physiol Psychol 66:25 – 34
Kennard MA (1954) The course of ascending fibers in the spinal cord of the cat essential to the recognition of painful stimuli. J Comp Neurol 100:511 – 524
Kenshalo DR, Jr. Leonard RB, Chung JM, Willis WD (1979) Responses of primate spinothalamic neurons to graded and to repeated noxious heat stimuli. J Neurophysiol 42:1370 – 1389
Kenshalo DR, Jr, Leonard RB, Chung JM, Willis WD (1982) Facilitation of the responses of primate spinothalamic cells to cold and to tactile stimuli by noxious heating of the skin. Pain 12:141 – 152
Kerr FWL (1975) The ventral spinothalamic tract and other ascending systems of the ventral funiculus of the spinal cord. J Comp Neurol 159:335 – 356
Kerr FWL, Wilson PR (1978) Pain. Annu Rev Neurosci 1:83 – 102
Kerr FWL, Wilson PR, Nijensohn DE (1978) Acupuncture reduces the trigeminal evoked response in decerebrate cats. Exp Neurol 61:84 – 95
Kidokoro Y, Kubota K, Shuto S, Sumino R (1968) Reflex organization of cat masticatory muscles. J Neurophysiol 31:695 – 708
Kin KC, Han YF, Yu LP, Feng J, Wang FS, Zhang ZD, Chang AZ, Sheng MP, Lu YY (1979) Role of brain serotonergic and catecholaminergic systems in acupuncture analgesia. Acta Physiol Sin 31:121 – 131
Kirkwood PA, Sears TA (1974) Monosynaptic excitation of motoneurones from secondary endings of muscle spindles. Nature 252:243 – 244
Kirkwood PA, Sears TA (1975) Monosynaptic excitation of motoneurones from muscle spindle secondary endings of intercostal and triceps surae muscles in the cat. J Physiol 245:64 – 66P
Kitchell RL (1962) Methodological considerations for assessment of pain perception in animals. In Keele CA, Smith R (eds) The assessment of pain in man and animals. S. Livingstone, London, pp 244 – 261
Klepac RK, Dowling J, Hauge G (1981) Sensitivity of the McGill pain questionnaire to intensity and quality of laboratory pain. Pain 10:199 – 207
Kleyntjens F, Koizumi K, McC Brooks C (1955) Stimulation of suprabulbar reticular formation. Arch Neurol Psychiat 73:425 – 438
Kneisley LW, Biber MP, LaVail JH (1978) A study of the origin of brain stem projections to monkey spinal cord using the retrograde transport method. Exp Neurol 60:116 – 139
Knibestöl M (1975) Stimulus-response functions of slowly adapting mechanoreceptors in the human glabrous skin area. J Physiol (Lond) 245:63 – 80
Kniffki DK, Mense S, Schmidt RF (1977) The spinocervical tract as a possible pathway for muscular nociception. J Physiol (Paris) 73:359 – 366

Kobayashi RM, Palkovits M, Miller RJ, Chang KJ, Cuatrecasas P (1978) Brain enkephalin distribution is unaltered by hypophysectomy. Life Sci 22:527 – 530
Kolmodin GM, Skoglund CR (1960) Analysis of spinal interneurons activated by tactile and nociceptive stimulation. Acta Physiol Scand 50:337 – 355
Komisaruk BR, Ciofalo B, Latranyi MB (1976) Stimulation of the vaginal cervix is more effective than morphine in suppressing a nociceptive response in rats. Adv Pain Res Ther 1:439 – 443
Komisaruk BR, Wallman J (1977) Antinociceptive effects of vaginal stimulation in rats: neurophysiological and behavioral studies. Brain Res 137:85 – 107
Kosterlitz HW (1978) Endogenous opioid peptides: historical aspects. In Hughes J (ed) Centrally acting peptides. Univ Park Press, Baltimore
Kosterlitz HW, Hughes J (1975) Some thoughts on the significance of enkephalin, the endogenous ligand. Life Sci 17:91 – 96
Kravitz EA, Kuffler SW, Potter DD (1963) Gamma-aminobutyric acid and other blocking compounds in crustacea. III. Their relative concentrations in separated motor and inhibitory axons. J Neurophysiol 26:739 – 751
Kremer E, Atkinson JH, Ignelzi RJ (1981) Measurement of pain: patient preference does not confound pain measurement. Pain 10:241 – 248
Kroger WS (1973) Acupunctural analgesia: its explanation by conditioning theory, autogenic training, and hypnosis. Am J Psychiatry 130:855 – 860
Kuffler SW, Eyzaguirre C (1955) Synaptic inhibition in an isolated nerve cell. J Gen Physiol 39:155 – 184
Kuhar MJ, Pert CB, Snyder SH (1973) Regional distribution of opiate receptor binding in monkey and human brain. Nature 245:447 – 450
Kumazawa T, Perl ER (1978) Excitation of marginal and substantia gelatinosa neurons in the primate spinal cord: indications of their place in dorsal horn functional organization. J Comp Neurol 177:417 – 434
Kumazawa T, Perl ER, Burgess PR, Whitehorn D (1975) Ascending projections from marginal zone (lamina I) neurons of the spinal dorsal horn. J Comp Neurol 162:1 – 12
Kuno M, Perl ER (1960) Alteration of spinal reflexes by interaction with suprasegmental and dorsal root activity. J Physiol (Lond) 151:103 – 122
Kuru M (1949) Sensory paths in the spinal cord and brain stem of man. Sogensya, Tokyo
Kuypers HGJM (1958) An anatomical analysis of cortico-bulbar connexions to the pons and lower brain stem in the cat. J Anat 92:198 – 218
Kuypers HGJM, Maisky VA (1975) Retrograde axonal transport of horseradish peroxidase from spinal cord to brain stem cell groups in the cat. Neurosci Lett 1:9 – 14
Kuypers HGJM, Maisky VA (1977) Funicular trajectories of descending brain stem pathways in cat. Brain Res 136:159 – 165
Kuypers HGJM, Tuerk JD (1964) The distribution of the cortical fibres within the nuclei cuneatus and gracilis in the cat. J Anat 98:143 – 162
Lal H, Spaulding T, Fielding S (1978) Swim-stress induced analgesia and lack of its naloxone antagonism. Commun Psychopharmacol 2:263 – 266
LaMotte C, Pert CB, Snyder SH (1976) Opiate receptor binding in primate spinal cord: distribution and changes after dorsal root section. Brain Res 112:407 – 412
Laporte Y, Boër A (1954) Décharge réflexe segmentaire homolatérale provoquée par les fibres amyéliniques chez le Chat spinal aigu. C R Soc Biol (Paris) 148:793 – 796
Laporte Y, Boër A (1955) Réflexe de flexion ipsilatéral provoquée par les fibres amyéliniques chez le Chat spinal. C R Soc Biol (Paris) 149:808 – 810
Laporte Y, Bessou P (1958) Réflexes ipsi-latéraux d'origine exclusivement amyélinique chez le Chat. C R Soc Biol (Paris) 52:161 – 164
Laporte Y, Lloyd DPC (1952) Nature and significance of the reflex connections established by large afferent fibers of muscular origin. Am J Physiol 169:609 – 621

Larsson LI, Childers S, Snyder SH (1979) Met- and leu-enkephalin immunoreactivity in separate neurones. Nature 282:407–410
Lazarus LH, Ling N, Guillemin R (1976) β-Lipotropin as a prohormone for the morphinomimetic peptides endorphins and enkephalins. Proc Natl Acad Sci USA 73:2156–2159
LeBars D, Guilbaud G, Jurna I, Besson JM (1976a) Differential effects of morphine on responses of dorsal horn lamina V type cells elicited by A and C fibre stimulation in the spinal cat. Brain Res 115:518–524
LeBars D, Menétrey D, Besson JM (1976b) Effects of morphine upon the lamina V type cells activities in the dorsal horn of the decerebrate cat. Brain Res 113:293–310
LeBars D, Dickenson AH, Besson JM (1979a) Diffuse noxious inhibitory controls (DNIC). I. Effects on dorsal horn convergent neurones in the rat. Pain 6:283–304
LeBars D, Dickenson AH, Besson JM (1979b) Diffuse noxious inhibitory controls (DNIC). I. Lack of effect on non-convergent neurones, supraspinal involvement and theoretical implications. Pain 6:305–327
LeBars D, Rivot JP, Guilbaud G, Menétrey D, Besson JM (1979c) The depressive effect of morphine on the C fibre response of dorsal horn neurones in the spinal rat pretreated or not by *p*CPA. Brain Res 176:337–353
LeBars D, Dickenson AH, Besson JM (1980) Microinjection of morphine within nucleus raphe magnus and dorsal horn neurone activities related to nociception in the rat. Brain Res 189:467–481
LeBars D, Chitour D, Kraus E, Dickenson AH, Besson JM (1981a) Effect of naloxone upon diffuse noxious inhibitory controls (DNIC) in the rat. Brain Res 204:387–402
LeBars D, Chitour D, Kraus E, Clot AM, Dickenson AH, Besson JM (1981b) The effect of systemic morphine upon diffuse noxious inhibitory controls (DNIC) in the rat: evidence for a lifting of certain descending inhibitory controls of dorsal horn convergent neurones. Brain Res 215:257–274
Lee GTC, Yang MMP, Kok SH (1976) Acupuncture anesthesia used in rabbit abdominal operations. Am J Acupunc 4:149–156
Lee PK, Andersen TW, Modell JH, Saga SA (1975) Treatment of chronic pain with acupuncture. JAMA 232:1133–1135
Lee PKY, Modell JH, Andersen TW, Saga SA (1976) Incidence of prolonged pain relief following acupuncture. Anesth Analg (Cleve) 55:229–231
Leichnetz GL, Watkins L, Griffin G, Murfin R, Mayer DJ (1978) The projection from the nucleus raphe magnus and other brainstem nuclei to the spinal cord in the rat: a study using the HRP blue-reaction. Neurosci Lett 8:119–124
Leksell L (1945) The action potential and excitatory effects of the small ventral root fibres to skeletal muscle. Acta Physiol Scand [Suppl 31] 10:1–84
Levante A, Lamour Y, Guilbaud G, Besson JM (1975) Spinothalamic cell activity in the monkey during intense nociceptive stimulation: intra-arterial injection of bradykinin into the limbs. Brain Res 88:560–564
Levine JD, Gordon NC, Jones RT, Fields HL (1978) The narcotic antagonist naloxone enhances clinical pain. Nature 272:826–827
Levitt EE, Walker FD (1975) Evaluation of acupuncture in treatment of chronic pain. J Chronic Dis 28:311–316
Levitt M, Carreras M, Liu CN, Chambers WW (1964) Pyramidal and extrapyramidal modulation of somatosensory activity in gracile and cuneate nuclei. Arch Ital Biol 102:197–229
Lewis JW, Cannon JT, Liebeskind JC (1980) Opioid and nonopioid mechanisms of stress analgesia. Science 208:623–625
Lewis T (1942) Pain. Macmillan, New York
Lewis VA, Gebhart GF (1977a) Evaluation of the periaqueductal central gray (PAG) as a morphine-specific locus of action and examination of morphine-induced and stimulation-produced analgesia at coincident PAG loci. Brain Res 124:283–303

Lewis VA, Gebhart GF (1977b) Morphine-induced and stimulation-produced analgesias at coincident periaqueductal central gray loci: evaluation of analgesic congruence, tolerance and cross-tolerance. Exp Neurol 57:934 – 955

Li CH, Chung D (1976) Isolation and structure of an untriakontapeptide with opiate activity from camel pituitary glands. Proc Natl Acad Sci USA 73:1145 – 1148

Li CH, Chung D, Doneen BA (1976) Isolation, characterization and opiate activity of β-endorphin from human pituitary glands. Biochem Biophys Res Commun 72:1542 – 1547

Li CL (1973) Neurological basis of pain and its possible relationship to acupuncture-analgesia. Am J Chin Med 1:61 – 72

Liddell EGT, Matthes K, Oldberg E, Ruch TC (1932) Reflex release of flexor muscles by spinal section. Brain 55:239 – 246

Liddell EGT, Sherrington CS (1923) A comparison between certain feactures of the spinal flexor reflex and of the decerebrate extensor reflex respectively. Proc R Soc Lond [Biol] 95:299 – 339

Liebeskind JC, Guilbaud G, Besson JM, Oliveras JL (1973) Analgesia from electrical stimulation of the periqueductal gray matter in the cat: behavioral observations and inhibitory effects on spinal cord interneurons. Brain Res 50:441 – 446

Light AR, Trevino DL, Perl ER (1979) Morphological features of functionally defined neurons in the marginal zone and substantia gelatinosa of the spinal dorsal horn. J Comp Neurol 186:151 – 172

Lindberg I, Smythe SJ, Dahl JL (1979) Regional distribution of enkephalin in bovine brain. Brain Res 168:200 – 204

Lindblom U, Meyerson BA (1975) Influence on touch, vibration and cutaneous pain of dorsal column stimulation in man. Pain 1:257 – 270

Lindblom UF, Ottosson JO (1953) Effects of spinal sections on the spinal cord potentials elicited by stimulation of low threshold cutaneous fibres. Acta Physiol Scand [Suppl 106] 29:191 – 208

Lindblom UF, Ottosson JO (1956) Bulbar influence on spinal cord dorsum potentials and ventral root reflexes. Acta Physiol Scand 35:203 – 214

Lindblom U, Tegnér R (1979) Are the endorphins active in clinical pain states? Narcotic antagonism in chronic pain patients. Pain 7:65 – 68

Lineberry CG, Vierck CJ (1975) Attenuation of pain reactivity by caudate nucleus stimulation in monkeys. Brain Res 98:119 – 134

Livingston RB (1959) Central control of receptors and sensory transmission systems. In: Magoun HW (ed) Neurophysiology, vol 1. Washington, Amer Physiol Soc (Handbook of physiology, sec. 1, pp 741 – 760)

Llinas R, Terzuolo CA (1964) Mechanisms of supraspinal actions upon spinal cord activities. Reticular inhibitory mechanisms on alpha-extensor motoneurons. J Neurophysiol 27:579 – 591

Lloyd DPC (1943a) Reflex action in relation to the pattern and peripheral source of afferent stimulation. J Neurophysiol 6:111 – 119

Lloyd DPC (1943b) Neuron patterns controlling transmission of ipsilateral hindlimb reflexes in cat. J Neurophysiol 6:293 – 315

Lloyd DPC (1943c) Conduction and synaptic transmission of the reflex response to stretch in spinal cats. J Neurophysiol 6:317 – 326

Lloyd DPC (1946a) Facilitation and inhibition of spinal motoneurons. J Neurophysiol 9:421 – 438

Lloyd DPC (1946b) Integrative pattern of excitation and inhibition in two-neuron reflex arc. J Neurophysiol 9:439 – 444

Lloyd DPC (1957) Input-output relation in a flexor reflex. J Gen Physiol 41:297 – 306

Lloyd DPC, Chang HT (1948) Afferent fibers in muscle nerves. J Neurophysiol 11:199 – 207

Lloyd MA, Wagner MK (1976) Acupuncture analgesia and radiant-heat pain: a signal detection analysis. Anesthesiology 44:147–150

Loan WB, Morrison JD, Dundee JW (1968) Evaluation of a method for assessing potent analgesics. Clin Pharmacol Ther 9:765–776

Loeser JD, Black RG, Christman A (1975) Relief of pain by transcutaneous stimulation. J Neurosurg 42:308–314

Loh HH, Tseng LF, Wei E, Li CH (1976) β-Endorphin is a potent analgesic agent. Proc Natl Acad Sci USA 73:2895–2898

Long DM (1973) Electrical stimulation for relief of pain from chronic nerve injury. J Neurosurg 39:718–722

Long DM, Hagfors N (1975) Electrical stimulation in the nervous system: the current status of electrical stimulation of the nervous system for the relief of pain. Pain 1:109–123

Lord JAH, Waterfield AA, Hughes J, Kosterlitz HW (1977) Endogenous opioid peptides: Multiple agonists and receptors. Nature 267:495–499

Lovick TA, West DC, Wolstencroft JH (1978) Responses of raphespinal and other bulbar raphe neurones to stimulation of the periaqueductal gray in the cat. Neurosci Lett 8:45–49

Lu GD, Needham J (1980) Celestial lancets. A history and rationale of acupuncture and moxa. Cambridge University Press, Cambridge

Lundberg A, Oscarsson O (1961) Three ascending spinal pathways in the dorsal part of the lateral funiculus. Acta Physiol Scand 51:1–16

Lundberg A, Oscarsson O (1962) Two ascending spinal pathways in the ventral part of the cord. Acta Physiol Scand 51:270–286

Lundberg A, Malmgren K, Schomburg ED (1977) Comments on reflex actions evoked by electrical stimulation of group II muscle afferents. Brain Res 122:551–555

Lundberg A, Malmgren K, Schomburg ED (1978) Role of joint afferents in motor control exemplified by effects on reflex pathways from Ib afferents. J Physiol (Lond) 284:327–343

Lynn B, Perl ER (1977) Failure of acupuncture to produce localized analgesia. Pain 3:339–351

Madden J, Akil H, Patrick RL, Barchas JD (1977) Stress-induced parallel changes in central opioid levels and pain responsiveness in the rat. Nature 265:358–360

Magoun HW, Rhines R (1946) An inhibitory mechanism in the bulbar reticular formation. J Neurophysiol 9:165–171

Mains RE, Eipper BA, Ling N (1977) Common precursor to corticotropins and endorphins. Proc Natl Acad Sci USA 74:3014–3018

Major CT, Pleuvry BJ (1971) Effects of α-methyl-*p*-tyrosine, *p*-chlorophenylalanine, L-β-(3,4-dihydroxyphenyl)alanine, 5-hydroxytryptophan and diethyldithiocarbamate on the analgesic activity of morphine and methylamphetamine in the mouse. Br J Pharmacol 42:512–521

Malick JB, Goldstein JM (1977) Analgesic activity of enkephalins following intracerebral administration in the rat. Life Sci 20:827–832

Mann F (1972) Acupuncture in dentistry. Lancet 1:898–899

Mann F (1974) Acupuncture analgesia: report of 100 experiments. Br J Anaesthesia 46:361–364

Mann F, Bowsher D, Mumford J, Lipton S, Miles J (1973) Treatment of intractable pain by acupuncture. Lancet 2:57–60

Mannheimer C, Carlsson CA (1979) The analgesic effect of transcutaneous electrical nerve stimulation (TNS) in patients with rheumatoid arthritis. A comparative study of different pulse patterns. Pain 6:329–334

Margalit D, Segal M (1979) A pharmacologic study of analgesia produced by stimulation of the nucleus locus coeruleus. Psychopharmacology 62:169–173

Mark RF, Steiner J (1958) Cortical projection of impulses in myelinated cutaneous afferent nerve fibres of the cat. J Physiol 142:544 – 562
Markowitz R, Jacobson J, Bain C, Kornetsky C (1976) Naloxone blockade of morphine analgesia: a dose-effect study of duration and magnitude. J Pharmacol Exp Ther 199:385 – 388
Marmarelis PZ, Naka KI (1973) Nonlinear analysis and synthesis of receptive-field responses in the catfish retina. I. Horizontal cell-ganglion cell chain. J Neurophysiol 36:605 – 618
Martin GF, Humbertson AO, Laxson LC, Panneton WM, Tshismadia I (1979) Spinal projections from the mesencephalic and pontine reticular formation in the North American opossum: a study using axonal transport techniques. J Comp Neurol 187:373 – 400
Martin RF, Haber LH, Willis WD (1979) Primary afferent depolarization of identified cutaneous fibers following stimulation in medial brain stem. J Neurophysiol 42:779 – 790
Martin RF, Jordan LM, Willis WD (1978) Differential projections of cat medullary raphe neurons demonstrated by retrograde labelling following spinal cord lesions. J Comp Neurol 182:77 – 88
Martin RW, Chapman CR (1979) Dental dolorimetry for human pain research: Methods and apparatus. Pain 6:349 – 364
Martin WR (1967) Opioid antagonists. Pharmacol Rev 19:463 – 521
Martin WR, Eades CG, Fraser HF, Wikler A (1964) Use of hindlimb reflexes of the chronic spinal dog for comparing analgesics. J Pharmacol Exp Ther 144:8 – 11
Martin WR, Eades CG, Thompson JA, Huppler RE, Gilbert PE (1976) The effects of morphine- and nalorphine-like drugs in the nondependent and morphine-dependent chronic spinal dog. J Pharmacol Exp Ther 197:517 – 532
Matthews MA (1968) An electron microscopic study of the relationship between axon diameter and the initiation of myelin production in the peripheral nervous system. Anat Rec 161:337 – 352
Matthews PBC (1969) Evidence that the secondary as well as the primary endings of the muscle spindles may be responsible for the tonic stretch reflex of the decerebrate cat. J Physiol (Lond) 204:365 – 393
Matthews PBC (1972) Mammalian muscle receptors and their central actions. Arnold, London
Matthews PBC (1973) A critique of the hypothesis that the spindle secondary endings contribute excitation to the stretch reflex. In: Stein RB, Pearson KS, Smith RS, Redford JB (eds) Control of posture and locomotion. Plenum Press, New York, pp 227 – 243
Mayer DJ (1975) Pain inhibition by electrical brain stimulation: comparison to morphine. Neurosci Res Program Bull 13:94 – 99
Mayer DJ, Hayes RL (1975) Stimulation produced analgesia: development of tolerance and cross-tolerance to morphine. Science 188:941 – 943
Mayer DJ, Liebeskind JC (1974) Pain reduction by focal electrical stimulation of the brain: an anatomical and behavioral analysis. Brain Res 68:73 – 93
Mayer DJ, Price DD (1976) Central nervous system mechanisms of analgesia. Pain 2:379 – 404
Mayer DJ, Wolfle TL, Akil H, Carder B, Liebeskind JC (1971) Analgesia from electrical stimulation in the brainstem of the rat. Science 174:1351 – 1354
Mayer DJ, Price DD, Becker DP (1975) Neurophysiological characterization of the anterolateral spinal cord neurons contributing to pain perception in man. Pain 1:51 – 58
Mayer DJ, Price DD, Rafii A (1977) Antagonism of acupuncture analgesia in man by the narcotic antagonist naloxone. Brain Res 121:368 – 372
Mazars GL (1975) Intermittent stimulation of nucleus ventralis posterolateralis for intractable pain. Surg Neurol 4:93 – 94

Mazars G, Merienne L, Cioloca C (1973) Stimulations thalamiques intermittentes antalgiques. Rev Neurol 128:273 – 279
Mazars GJ, Merienne L, Cioloca C (1979) Comparative study of electrical stimulation of posterior thalamic nuclei, periaqueductal gray, and other midline mesencephalic structures in man. Adv Pain Res Ther 3:541 – 546
McBurney DH (1975) Acupuncture, pain, and signal detection theory. Science 189:66
McBurney DH (1976) Signal detection theory and pain. Anesthesiology 44:356 – 358
McCreery DB, Bloedel JR (1975) Reduction of the response of cat spinothalamic neurons to graded mechanical stimuli by electrical stimulation of the lower brain stem. Brain Res 97:151 – 156
McCreery DB, Bloedel JR (1976) Effect of trigeminal stimulation on the excitability of cat spinothalamic neurons. Brain Res 117:136 – 140
McCreery DB, Bloedel JR (1978) A critical examination of the use of signal detection theory in evaluating a putative analgesic-transcutaneous electrical nerve stimulation. Sens Processes 2:38 – 57
McCreery DB, Bloedel JR, Hames EG (1979a) Excitability changes in lumbosacral spinothalamic neurons produced by non-noxious mechanical stimuli and by graded electrical stimuli applied to the face. Brain Res 177:253 – 263
McCreery DB, Bloedel JR, Hames EG (1979b) Effects of stimulating in raphe nuclei and in reticular formation on response of spinothalamic neurons to mechanical stimuli. J Neurophysiol 42:166 – 182
McDonald RL, Nelson PG (1978) Specific-opiate-induced depression of transmitter release from dorsal root ganglion cells in culture. Science 199:1449 – 1451
McGrath PA, Sharav Y, Dubner R, Gracely RH (1981) Masseter inhibitory periods and sensations evoked by electrical tooth pulp stimulation. Pain 10:1 – 17
McIntyre AK (1974) Central actions of impulses in muscle afferent fibers. In: Hunt CC (ed) Muscle receptors, Springer, Berlin Heidelberg New York (Handbook of sensory physiology, vol 3/2, pp 235 – 288)
McLennan H, Gilfillan K, Heap Y (1977) Some pharmacological observations on the analgesia induced by acupuncture in rabbits. Pain 3:229 – 238
Medakovič M, Banič B (1964) The action of reserpine and α-methyl-*m*-tyrosine on the analgesic effect of morphine in rats and mice. J Pharm Pharmacol 16:198 – 206
Mehler WR, Feferman ME, Nauta WJH (1960) Ascending axon degeneration following anterolateral cordotomy. An experimental study in the monkey. Brain 83:718 – 751
Melzack R (1973) The puzzle of pain. Basic Books, New York
Melzack R (1975a) The McGill pain questionnaire: Major properties and scoring methods. Pain 1:277 – 299
Melzack R (1975b) Prolonged relief of pain by brief, intense transcutaneous somatic stimulation. Pain 1:357 – 373
Melzack R, Stillwell DM, Fox EJ (1977) Trigger points and acupuncture points for pain : Correlations and implications. Pain 3:3 – 23
Melzack R, Stotler WA, Livingston WK (1958) Effects of discrete brainstem lesions in cats on perception of noxious stimulation. J Neurophysiol 21:353 – 367
Melzack R, Torgerson WS (1971) On the language of pain. Anesthesiology 34:50 – 59
Melzack R, Melinkoff DF (1974) Analgesia produced by brain stimulation: evidence of a prolonged onset period. Exp Neurol 43:369 – 374
Mendell LM (1966) Physiological properties of unmyelinated fiber projection to the spinal cord. Exp Neurol 16:316 – 332
Menétrey D, Giesler GJ, Besson JM (1977) An analysis of response properties of spinal cord dorsal horn neurones to nonnoxious and noxious stimuli in the spinal rat. Exp Brain Res 27:15 – 33
Mense S (1977) Nervous outflow from skeletal muscle following chemical noxious stimulation. J Physiol (Lond) 267:75 – 88

Messing RB, Lytle LD (1977) Serotonin-containing neurons: their possible role in pain and analgesia. Pain 4:1 – 22
Messing RB, Phebus L, Fisher LA, Lytle LD (1975) Analgesic effect of fluoxetine hydrochloride (Lilly 110140), a specific inhibitor of serotonin uptake. Psychopharm Commun 1:511 – 521
Messing RB, Fisher LA, Phebus L, Lytle LD (1976) Interaction of diet and drugs in the regulation of brain 5-hydroxyindoles and the response to painful electric shock. Life Sci 18:707 – 714
Meyer GA, Fields HL (1972) Causalgia treated by selective large fibre stimulation of peripheral nerve. Brain 95:163 – 168
Miles J, Lipton S (1978) Phantom limb pain treated by electrical stimulation. Pain 5:373 – 382
Millan MJ, Przewlocki R, Herz A (1980) A non-β-endorphinergic adenohypophyseal mechanism is essential for an analgetic response to stress. Pain 8:343 – 353
Milne RJ, Foreman RD, Giesler GJ, Willis WD (1981) Convergence of cutaneous and pelvic visceral nociceptive inputs onto primate spinothalamic neurons. Pain 11:163 – 183
Mitchell CL (1964) A comparison of drug effects upon the jaw jerk to electrical stimulation of the tooth pulp in dogs and cats. J Pharmacol Exp Ther 146:1 – 6
Modell JH, Lee PKY, Bingham HG, Greer DM, Habal MB (1976) "Acupuncture anesthesia" a clinical study. Anesth Analg (Cleve) 55:508 – 512
Mohrland JS, Gebhart GF (1980) Effects of focal electrical stimulation and morphine microinjection in the periaqueductal gray of the rat mesencephalon and neuronal activity in the medullary reticular formation. Brain Res 201:23 – 37
Moolenaar GM, Holloway JA, Trouth CO (1976) Responses of caudal raphe neurons to peripheral somatic stimulation. Exp Neurol 53:304 – 313
Moore PA, Duncan GH, Scott DS, Gregg JM, Ghia JN (1979) The submaximal effort tourniquet test: its use in evaluating experimental and chronic pain. Pain 6:375 – 382
Mudge AW, Leemann SE, Fischbach GD (1979) Enkephalin inhibits release of substance P from sensory neurons in cultures and decreases action potential duration. Proc Natl Acad Sci USA 76:526 – 530
Mumford J, Bowsher D (1973) Electro-acupuncture and pain threshold. Lancet 2:667
Naka KI, Rushton WAH (1966) S-potentials from colour units in the retina of fish (Cyprinidae). J Physiol (Lond) 185:536 – 555
Nakamura K, Kuntzman R, Maggio AC, Augulis V, Conney AH (1973) Influence of 6-hydroxydopamine on the effect of morphine on the tail-flick latency. Psychopharmacologia 31:177 – 189
Nashold BS, Friedman H (1972) Dorsal column stimulation for control of pain. Preliminary report on 30 patients. J Neurosurg 36:590 – 597
Nashold BS, Wilson WP, Slaughter DG (1969) Sensations evoked by stimulation in the midbrain of man. J Neurosurg 30:14 – 24
National Symposia of Acupuncture and Moxibustion and Acupuncture Anaesthesia (1979) China, Beijing
Nghi N, Fisch G, Kao J (1973) An introduction to classical acupuncture. Am J Chin Med 1:75 – 83
Nielson KD, Adams JE, Hosobuchi Y (1975a) Phantom limb pain; treatment with dorsal column stimulation. J Neurosurg 42:301 – 307
Nielson KD, Adams JE, Hosobuchi Y (1975b) Experience with dorsal column stimulation for relief of chronic intractable pain 1968 – 1973. Surg Neurol 4:148 – 152
Ninkovic M, Hunt SP, Emson PC, Iversen LL (1981) The distribution of multiple opiate receptors in bovine brain. Brain Res 214:163 – 167
Noordenbos W, Wall PD (1976) Diverse sensory functions with an almost totally divided spinal cord. A case of spinal cord transection with preservation of part of one anterolateral quadrant. Pain 2:185 – 195

Nyberg-Hansen R (1965) Sites and mode of termination of reticulo-spinal fibers in the cat. An experimental study with silver impregnation methods. J Comp Neurol 124:71 – 100

O'Connor J, Benskey D (1975) A summary of research concerning the effects of acupuncture. Am J Chin Med 3:377 – 394

Ohnhaus EE, Adler R (1975) Methodological problems in the measurement of pain: a comparison between the verbal rating scale and the visual analogue scale. Pain 1:379 – 384

Oleson TD, Twombly DA, Liebeskind JC (1978) Effects of pain-attenuating brain stimulation and morphine on electrical activity in the raphe nuclei of the awake rat. Pain 4:211 – 230

Oleson TD, Kirkpatrick DB, Goodman SJ (1980) Elevation of pain threshold to tooth shock by brain stimulation in primates. Brain Res 194:79 – 95

Oliver C, Mical RS, Porter JC (1977) Hypothalamic-pituitary vasculature: Evidence for retrograde blood flow in the pituitary stalk. Endocrinology 101:598 – 604

Oliveras JL, Besson JM, Guilbaud G, Liebeskind JC (1974a) Behavioral and electrophysiological evidence of pain inhibition from midbrain stimulation in the cat. Exp Brain Res 20:32 – 44

Oliveras JL, Woda A, Guilbaud G, Besson JM (1974b) Inhibition of the jaw opening reflex by electrical stimulation of the periaqueductal gray matter in the awake, unrestrained cat. Brain Res 72:328 – 331

Oliveras JL, Redjemi F, Guilbaud G, Besson JM (1975) Analgesia induced by electrical stimulation of the inferior centralis nucleus of the raphe in the cat. Pain 1:139 – 145

Oliveras JL, Hosobuchi Y, Redjemi F, Guilbaud G, Besson JM (1977) Opiate antagonist, naloxone, strongly reduces analgesia induced by stimulation of a raphe nucleus (centralis inferior). Brain Res 120:221 – 229

Oliveras JL, Hosobuchi G, Guilbaud G, Besson JM (1978) Analgesic electrical stimulation of the feline nucleus raphe magnus: development of tolerance and its reversal by 5-HTP. Brain Res 146:404 – 409

Oliveras JL, Guilbaud G, Besson JM (1979) A map of serotoninergic structures involved in stimulation producing analgesia in unrestrained freely moving cats. Brain Res 164:317 – 322

Orne MT (1974) Pain suppression by hypnosis and related phenomena. Adv Neurol 4:563 – 572

Paintal AS (1960) Functional analysis of Group III afferent fibres of mammalian muscles. J Physiol (Lond) 152:250 – 270

Paintal AS (1961) Participation by pressure-pain receptors of mammalian muscles in the flexion reflex. J Physiol (Lond) 156:498 – 514

Pasternak GW, Goodman R, Snyder SH (1975) An endogenous morphine-like factor in mammalian brain. Life Sci 16:1765 – 1769

Peking Acupuncture Anaesthesia Co-ordinating Group (1973) Preliminary study on the mechanism of acupuncture anesthesia. Sci Sin 16:447 – 456

Perl ER (1957) Crossed reflexes of cutaneous origin. Am J Physiol 188:609 – 615

Perl ER (1958) Crossed reflex effects evoked by activity in myelinated afferent fibers of muscle. J Neurophysiol 21:101 – 112

Perl ER (1968) Myelinated afferent fibres innervating the primate skin and their response to noxious stimuli. J Physiol (Lond) 197:593 – 615

Peroutka SJ, Lebovitz RM, Snyder SH (1981) Two distinct central serotonin receptors with different physiological functions. Science 212:827 – 829

Peroutka SJ, Snyder SH (1979) Multiple serotonin receptors: differential binding of (^{3}H) 5-hydroxytryptamine, (^{3}H) lysergic acid diethylamide and (^{3}H) spiroperidol. Mol Pharmacol 16:687 – 699

Pert A, Yaksh T (1974) Sites of morphine induced analgesia in the primate brain: relation to pain pathways. Brain Res 80:135 – 140

Pert A, Yaksh T (1975) Localization of the antinociceptive action of morphine in primate brain. Pharmacol Biochem Behav 3:133 – 138

Pert A, Walter M (1976) Comparison between naloxone reversal of morphine and electrical stimulation induced analgesia in the rat mesencephalon. Life Sci 19:1023 – 1032

Pert A, Dionne R, Ng L, Bragin E, Moody TW, Pert CB (1981) Alterations in rat central nervous system endorphins following transauricular electroacupuncture. Brain Res 224:83 – 93

Pert CB, Snyder SH (1973) Opiate receptor: demonstration in nervous tissue. Science 179:1011 – 1014

Pert CB, Kuhar MJ, Snyder SH (1975) Autoradiographic localization of the opiate receptor in rat brain. Life Sci 16:1849 – 1854

Pertovaara A, Kemppainen P (1981) The influence of naloxone on dental pain threshold elevation produced by peripheral conditioning stimulation at high frequency. Brain Res 215:426 – 429

Petras JM (1967) Cortical, tectal and tegmental fiber connections in the spinal cord of the cat. Brain Res 6:275 – 324

Pfaffmann C (1939) Afferent impulses from the teeth due to pressure and noxious stimulation. J Physiol (Lond) 97:207 – 219

Picaza JA, Cannon BW, Hunter SE, Boyd AS, Guma J, Maurer D (1975) Pain suppression by peripheral nerve stimulation. I. Observations with transcutaneous stimuli. Surg Neurol 4:105 – 114

Pierce ET, Foote WE, Hobson JA (1976) The efferent connection of the nucleus raphe dorsalis. Brain Res 107:137 – 144

Pi-Suñer J, Fulton JF (1929) The influence of the proprioceptive system upon the crossed extensor reflex. Am J Physiol 88:453 – 467

Pomeranz B, Chiu D (1976) Naloxone blockade of acupuncture analgesia: endorphine implicated. Life Sci 19:1757 – 1762

Pomeranz B, Paley D (1979) Electroacupuncture hypalgesia is mediated by afferent nerve impulses: an electrophysiological study in mice. Exp Neurol 66:398 – 402

Pomeranz B, Wall PD, Weber WV (1968) Cord cells responding to fine myelinated afferents from viscera, muscle and skin. J Physiol (Lond) 199:511 – 532

Pomeranz B, Cheng R, Law P (1977) Acupuncture reduces electrophysiological and behavioral responses to noxious stimuli: pituitary is implicated. Exp Neurol 54:172 – 178

Pomeroy SL, Behbehani MM (1979) Physiologic evidence for a projection from periaqueductal gray to nucleus raphe magnus in the rat. Brain Res 176:143 – 147

Price DD (1972) Characteristics of second pain and flexion reflexes indicative of prolonged central summation. Exp Neurol 37:371 – 387

Price DD, Browe AC (1973) Responses of spinal cord neurons to graded noxious and non-noxious stimuli. Brain Res 64:425 – 429

Price DD, Mayer DJ (1974) Physiological laminar organization of the dorsal horn of *M. mulatta*. Brain Res 79:321 – 325

Price DD, Mayer, DJ (1975) Neurophysiological characterization of the anterolateral quadrant neurons subserving pain in *M. mulatta*. Pain 1:59 – 72

Price DD, Hayes RL, Ruda M, Dubner R (1978) Spatial and temporal transformations of input to spinothalamic tract neurons and their relation to somatic sensations. J Neurophysiol 41:933 – 947

Price DD, Hayashi H, Dubner R, Ruda MA (1979) Functional relationships between neurons of marginal and substantia gelatinosa layers of primate dorsal horn. J Neurophysiol 42:1590 – 1608

Price MTC, Fibiger HC (1975) Ascending catecholamine systems and morphine analgesia. Brain Res 99:189 – 193

Proudfit HK (1980) Reversible inactivation of raphe magnus neurons: effects on nociceptive threshold and morphine-induced analgesia. Brain Res 201:459–464

Proudfit HK, Anderson EG (1973) Influence of serotonin antagonists on bulbospinal systems. Brain Res 61:331–341

Proudfit HK, Anderson EG (1974) New long latency bulbospinal evoked potentials blocked by serotonin antagonists. Brain Res 65:542–546

Proudfit HK, Anderson EG (1975) Morphine analgesia: blockade by raphe magnus lesions. Brain Res 98:612–618

Proudfit HK, Levy RA (1978) Delimitation of neuronal substrates necessary for the analgesic action of baclofen and morphine. Eur J Pharmacol 47:159–166

Proudfit HK, Larson AA, Anderson EG (1980) The role of GABA and serotonin in the mediation of raphe-evoked spinal cord dorsal root potentials. Brain Res 195:149–165

Proudfit, HK, Hammond DL (1981) Alterations in nociceptive threshold and morphine-induced analgesia produced by intrathecally administered amine antagonists. Brain Res 218:393–399

Puig MM, Gascón P, Musacchio JM (1977) Endorphin release: cross tolerance to morphine. Eur J Pharmacol 45:205–206

Radouco-Thomas S, Radouco-Thomas C, LeBreton E (1957) Action de la noradrénaline et de la réserpine sur l'analgésie expérimentale. Arch Exp Pathol Pharmakol 232:279–281

Ranck JB (1975) Which elements are excited in electrical stimulation of mammalian central nervous system: a review. Brain Res 98:417–440

Randić M, Miletić V (1978) Depressant actions of methionine-enkephalin and somatostatin in cat dorsal horn neurones activated by noxious stimuli. Brain Res 152:196–202

Randić M, Yu HH (1976) Effects of 5-hydroxytryptamine and bradykinin in cat dorsal horn neurones activated by noxious stimuli. Brain Res 111:197–203

Rasmussen GL (1946) The olivary peduncle and other fiber projections of the superior olivary complex. J. Comp Neurol 84:141–219

Reddy SVR, Yaksh TL (1980) Spinal noradrenergic system mediates antinociception. Brain Res 189:391–401

Reddy SVR, Maderdrut JL, Yaksh TL (1980) Spinal cord pharmacology of adrenergic agonist-mediated antinociception. J Pharmacol Exp Ther 213:525–533

Reinhold K, Bläsig J, Herz A (1973) Changes in brain concentration of biogenic amines and the antinociceptive effect of morphine in rats. Naunyn-Schmiedebergs Arch Pharmacol 278:69–80

Renshaw B (1940) Activity in the simplest spinal reflex pathways. J Neurophysiol 3:373–387

Revill SI, Robinson JO, Rosen M, Hogg MIJ (1976) The reliability of a linear analogue for evaluating pain. Anaesthesia 31:1191–1198

Rexed B (1952) The cytoarchitectonic organization of the spinal cord in the cat. J Comp Neurol 96:415–496

Rexed B, Therman PO (1948) Calibre spectra of motor and sensory nerve fibres to flexor and extensor muscles. J Neurophysiol 11:133–139

Reynolds AK, Randall LO (1957) Morphine and allied drugs. University of Toronto, Toronto

Reynolds DV (1969) Surgery in the rat during electrical analgesia induced by focal brain stimulation. Science 164:444–445

Reynolds DV, Sjoberg AI (1971) Neuroelectric research. CC Thomas, Springfield

Rhodes DL, Liebeskind JC (1978) Analgesia from rostral brain stem stimulation in the rat. Brain Res 143:521–532

Richardson DE, Akil H (1977a) Pain reduction by electrical brain stimulation in man: Part 1: Acute administration in periaqueductal and periventricular sites. J Neurosurg 47:178–183

Richardson DE, Akil H (1977b) Pain reduction by electrical brain stimulation in man: Part 2: Chronic self-administration in the periventricular gray matter. J Neurosurg 47:184–194

Ritchie TC, Leonard RB (1982) Immunocytochemical demonstration of serotonergic cells, terminals and axons in the spinal cord of the stingray *Dasyatis sabina*. Brain Res 240:334–337

Rivot JP, Chaouch A, Besson JM (1979) The influence of naloxone on the C fiber response of dorsal horn neurons and their inhibitory control by raphe magnus stimulation. Brain Res 176:355–364

Rivot JP, Chaouch A, Besson JM (1980) Nucleus raphe magnus modulation of response of rat dorsal horn neurons to unmyelinated fiber inputs: partial involvement of serotonergic pathways. J Neurophysiol 44:1039–1057

Roemer D, Buescher HH, Hill RC, Pless R, Bauer W, Cardinaux F, Closse A, Hauser D, Huguenin R (1977) A synthetic enkephalin analogue with prolonged parenteral and oral analgesic activity. Nature 268:547–549

Rollman GB (1977) Signal detection theory measurement of pain: a review and critique. Pain 3:187–211

Rollman GB (1979) Signal detection theory pain measures: empirical validation studies and adaptation-level effects. Pain 6:9–21

Rosenberg ME (1970) Synaptic connexions of alpha extensor motoneurones with ipsilateral and contralateral cutaneous nerves. J Physiol (Lond) 207:231–255

Rosenberg M, Curtis L, Bourke DL (1978) Transcutaneous electrical nerve stimulation for the relief of postoperative pain. Pain 5:129–133

Ross CA, Armstrong DM, Ruggiero DA, Pickel VM, Joh TH, Reis DJ (1981) Adrenaline neurons in the rostral ventrolateral medulla innervate thoracic spinal cord: a combined immunocytochemical and retrograde transport demonstration. Neurosci Lett 25:257–262

Ross GS (1966) A technique to study pain in monkeys: effect of drugs and anatomic lesions. In: Knighton RS, Dumke PR (eds) Pain. Little, Brown & Co, Boston, pp 91–110

Ross JW, Ashford A (1967) The effect of reserpine and α-methyldopa on the analgesic action of morphine in the mouse. J Pharm Pharmacol 19:709–713

Ross S, Krugman AD, Lyerly SB, Clyde DJ (1962) Drugs and placebos: A model design. Psychol Rep 10:383–392

Rossier J, Audigier, Y, Ling N, Cros J, Udenfriend S (1980) Met-enkephalin-Arg^6-Phe^7, present in high amounts in brain or rat, cattle and man, is an opioid agonist. Nature 288:88–90

Rossier J, French E, Gros C, Minick S, Guillemin R, Bloom FE (1979) Adrenalectomy, dexamethasone or stress alters opioid peptide levels in rat anterior pituitary but not intermediate lobe or brain. Life Sci 25:2105–2112

Rossier J, French ED, Rivier C, Ling N, Guillemin R, Bloom FE (1977) Foot-shock induced stress increases beta-endorphin levels in blood but not brain. Nature 270:618–620

Rubinstein M, Stein S, Gerber LD, Udenfriend S (1977) Isolation and characterization of the opioid peptides from rat pituitary: β-lipotropin. Proc Natl Acad Sci USA 74:3052–3055

Ruda MA (1975) An autoradiographic study of the efferent projections of the midbrain central gray in the cat. Doctoral dissertation, University of Pennsylvania, Philadelphia

Ruda MA, Gobel S (1980) Ultrastructural characterization of axonal endings in the substantia gelatinosa which take up [^{3}H] serotonin. Brain Res 184:57–83

Rudzik AD, Mennear JN (1965) Antagonism of analgesics by amine-depleting agents. J Pharm Pharmacol 17:326–327

Saarnivaara L (1969a) Effect of 5-hydroxytryptamine on morphine analgesia in rabbits. Ann Med Exp Fenn 47:113–123
Saarnivaara L (1969b) Analgesic activity of some sympathetic drugs and their effect on morphine analgesia in rabbits. Ann Med Exp Fenn 47:180–190
Sagen J, Proudfit HK (1981) Hypoalgesia induced by blockage of noradrenergic projections to the raphe magnus: reversal by blockade of noradrenergic projections to the spinal cord. Brain Res 223:391–396
Sakai K, Salvert D, Touret M, Jouvet M (1977) Afferent connections of the nucleus raphe dorsalis in the cat as visualized by the horseradish peroxidase technique. Brain Res 137:11–35
Salar G, Job I, Mingrino S, Bosio A, Trabucchi M (1981) Effect of transcutaneous electrotherapy on CSF β-endorphin content in patients without pain problems. Pain 10:169–172
Samanin R, Bernasconi S (1972) Effects of intraventricularly injected 6-OH dopamine or midbrain raphe lesion on morphine analgesia in rats. Psychopharmacologia 25:175–182
Samanin R, Gumulka W, Valzelli L (1970) Reduced effect of morphine in midbrain raphe lesioned rats. Eur J Pharmacol 10:339–343
Samanin R, Valzelli L (1971) Increase of morphine-induced analgesia by stimulation of the nucleus raphe dorsalis. Eur J Pharmacol 16:298–302
Sandberg DE, Segal M (1978) Pharmacological analysis of analgesia and self-stimulation elicited by electrical stimulation of catecholamine nuclei in the rat brain. Brain Res 152:529–542
Sanders KH, Klein CE, Mayer TE, Heym C, Handwerker HO (1980) Differential effects of noxious and non-noxious input on neurones according to location in ventral periaqueductal grey or dorsal raphe nucleus. Brain Res 186:83–97
Sar M, Stumpf WE, Miller RJ, Chang KJ, Cuatrecasas P (1978) Immunohistochemical localization of enkephalin in rat brain and spinal cord. J Comp Neurol 182:17–38
Sastry BR (1978) Morphine and met-enkephalin effects on sural Aδ afferent terminal excitability. Eur J Pharmacol 50:269–273
Sastry BR (1979) Presynaptic effects of morphine and methionine-enkephalin in feline spinal cord. Neuropharmacol 18:367–375
Satoh M, Takagi H (1971) Enhancement by morphine of the central descending inhibitory influence on spinal sensory transmission. Eur J Pharmacol 14:60–65
Satoh M, Nakamura N, Takagi H (1971) Effects of morphine on bradykinin-induced unitary discharges in the spinal cord of the rabbit. Eur J Pharmacol 16:245–247
Satoh M, Akaike A, Nakazawa T, Takagi H (1980a) Evidence for the involvement of separate mechanisms in the production of analgesia by electrical stimulation of the nucleus reticularis paragigantocellularis and nucleus raphe magnus in the rat. Brain Res 194:525–529
Satoh M, Kawajiri S, Yamamoto M, Akaike A, Ukai Y, Takagi H (1980b) Effects of tyrosyl-arginine (Kyotorphin), a new opioid dipeptide, on single neurons in the spinal cord dorsal horn of rabbits and the nucleus reticularis paragigantocellularis of rats. Neurosci Lett 16:319–322
Satran R, Goldstein MN (1973) Pain perception: modification of threshold of intolerance and cortical potentials by cutaneous stimulation. Science 180:1201–1202
Schaumann W (1958) Beeinflussung der analgetischen Wirkung des Morphins durch Reserpin. Naunyn-Schmiedebergs Arch Exp Pathol Pharmakol 235:1–9
Schmidek HH, Fohanno D, Ervin ER, Sweet WH (1971) Pain threshold alterations by brain stimulation in the monkey. J Neurosurg 35:715–722
Schmidt RF (1971) Presynaptic inhibition in the vertebrate central nervous system. Erg Physiol 63:20–101

Schmidt RF (1973) Control of the access of afferent activity to somatosensory pathways. In: Iggo A (ed) Somatosensory system. Springer, Berlin Heidelberg New York (Handbook of sensory physiology, vol 2, pp 151 – 206)
Schmitt P, Eclancher F, Karil P (1974) Etude des systèmes de renforcement négatif et de renforcement positif au niveau de la substance grise centrale chez le rat. Physiol Behav 12:271 – 279
Schuster GD, Infante MC (1980) Pain relief after low back surgery: the efficacy of transcutaneous electrical nerve stimulation. Pain 8:299 – 302
Segal M, Sandberg D (1977) Analgesia produced by electrical stimulation of catecholamine nuclei in the rat brain. Brain Res 123:369 – 372
Sessle BJ, Hu JW (1981) Raphe-induced suppression of the jaw-opening reflex and single neurons in trigeminal oralis, and influence of naloxone and subnucleus caudalis. Pain 10:19 – 36
Sessle BJ, Schmitt A (1972) Effects of controlled tooth stimulation on jaw muscle activity in man. Arch Oral Biol 17:1597 – 1607
Sewell RDE, Spencer PSJ (1976) Antinociceptive activity of narcotic agonist and partial agonist analgesics and other agents in the tail-immersion test in mice and rats. Neuropharmacol 15:683 – 688
Shah Y, Dostrovsky JO (1980) Electrophysiological evidence for a projection of the periaqueductal gray matter to nucleus raphe magnus in cat and rat. Brain Res 193:534 – 538
Sharpe LG, Garnett JE, Cicero TJ (1974) Analgesia and hyperreactivity produced by intracranial microinjections of morphine into the periaqueductal gray matter of the rat. Behav Biol 11:303 – 314
Shealy CN (1974) Transcutaneous electrical stimulation for control of pain. Clin Neurosurg 21:269 – 277
Shealy CN, Maurer D (1974) Transcutaneous nerve stimulation for control of pain. Surg Neurol 2:45 – 47
Shealy CN, Mortimer JT, Hagfors NR (1970) Dorsal column electroanalgesia. J Neurosurg 32:560 – 564
Sherman JE, Liebeskind JC (1980) An endorphinergic, centrifugal substrate of pain modulation: recent findings, current concepts, and complexities. In: Bonica JJ (ed) Pain. Raven Press, New York, pp 191 – 204
Sherrington CS (1903) Qualitative difference of spinal reflex corresponding with qualitative difference of cutaneous stimulus. J Physiol 30:39 – 46
Sherrington CS (1906) The integrative action of the nervous system. Yale University Press, New Haven
Sherrington CS (1910) Flexion-reflex of the limb, crossed extension-reflex and reflex stepping and standing. J Physiol (Lond) 40:28 – 121
Sherrington CS (1917) Reflexes elicitable in the cat from pinna, vibrissae and jaws. J Physiol (Lond) 51:404 – 431
Sherrington CS, Sowton SCM (1915) Observations on reflex responses to single break-shocks. J Physiol (Lond) 49:331 – 348
Shibuya T, Anderson EG (1968) The influence of chronic cord transection on the effects of 5-hydroxytryptophan, L-tryptophan and pargyline on spinal neuronal activity. J Pharmacol Exp Ther 164:185 – 190
Siegel P, Wepsic JG (1974) Alteration of nociception by stimulation of cerebellar structures in the monkey. Physiol Behav 13:189 – 194
Siegmund E, Cadmus R, Lu G (1957) A method for evaluating both non-narcotic and narcotic analgesics. Proc Sor Exp Biol Med 95:729 – 731
Simantov R, Snyder SH (1976a) Morphine-like peptides in mammalian brain: isolation, structure elucidation and interactions with the opiate receptor. Proc Natl Acad Sci USA 73:2515 – 2519

Simantov R, Snyder SH (1976b) Morphine-like peptides, leucine enkephalin and methionine enkephalin: Interactions with the opiate receptor. Mol Pharmacol 12:987–998

Simantov R, Snyder SH (1976c) Isolation and structure identification of a morphine-like peptide "enkephalin" in bovine brain. Life Sci 18:781–788

Simantov R, Kuhar MJ, Pasternak GW, Snyder SH (1976) The regional distribution of a morphine-like factor enkephalin in monkey brain. Brain Res 106:189–197

Simantov R, Kuhar MJ, Uhl GR, Snyder SH (1977) Opioid peptide enkephalin: immunohistochemical mapping in rat central nervous system. Proc Natl Acad Sci USA 74:2167–2171

Simon EJ, Hiller JM, Edelman I (1973) Stereospecific binding of the potent narcotic analgesic [^{3}H] etorphine to rat-brain homogenate. Proc Natl Acad Sci USA 70:1947–1949

Sjölund B, Eriksson M (1976) Electro-acupuncture and endogenous morphines. Lancet 2:1085

Sjölund BH, Eriksson MBE (1979) The influence of naloxone on analgesia produced by peripheral conditioning stimulation. Brain Res 173:295–301

Sjölund B, Terenius L, Eriksson M (1977) Increased cerebrospinal fluid levels of endorphines after electro-acupuncture. Acta Physiol Scand 100:382–384

Skoglund S (1956) Anatomical and physiological studies of knee joint innervation in the cat. Acta Physiol Scand [Suppl 124] 36:1–101

Skultety FM (1958) The behavioral effects of destructive lesions of the periaqueductal gray matter in adult cats. J Comp Neurol 110:337–365

Skultety FM (1963) Stimulation of periaqueductal gray and hypothalamus. Arch Neurol 8:608–620

Smith GM, Egbert LD, Markovitz RA, Mosteller F, Beecher HK (1966) An experimental pain method sensitive to morphine in man: the submaximum effort tourniquet technique. J Pharmacol Exp Ther 154:324–332

Snyder SH, Goodman RR (1980) Multiple neurotransmitter receptors. J Neurochem. 35:5–15

Sodipo JOA (1979) Therapeutic acupuncture for chronic pain. Pain 7:359–366

Soper WY (1976) Effects of analgesic midbrain stimulation on reflex withdrawal and thermal escape in the rat. J Comp Physiol Psychol 90:91–101

Sotgiu ML, Margnelli M (1976) Electrophysiological identification of pontomedullary reticular neurons directly projecting into dorsal column nuclei. Brain Res 103:443–453

Sotgiu ML, Marini G (1977) Reticulo-cuneate projections as revealed by horseradish peroxidase axonal transport. Brain Res 128:341–345

Sperry RW (1950) Neural basis of the spontaneous optokinetic response produced by visual inversion. J Comp Physiol Psychol 43:482–489

Spiaggia A, Bodnar RJ, Kelly DD, Glusman M (1979) Opiate and non-opiate mechanisms of stress-induced analgesia: cross-tolerance between stressors. Pharm Biochem Behav 10:761–765

Spiler IJ, Molitch ME (1980) Lack of modulation of pituitary hormone stress response by neural pathways involving opiate receptors. J Clin Endo Metab 50:516–520

Spoerel WE, Varkey M, Leung CY (1976) Acupuncture in chronic pain. Am J Chin Med 4:267–279

Stacher G, Bauer P, Steinringer H, Schreiber E, Schmierer G (1979) Effects of the synthetic enkephalin analogue FK 33-824 on pain threshold and pain tolerance in man. Pain 7:159–172

Starkenstein E (1932) Zur Methodik der Prüfung schmerzstillender Mittel im Tierversuch. Naunyn Schmiedeberg Arch Exp Pathol Pharmakol 165:325–338

Stauffer EK, Watt DGD, Taylor A, Reinking RM, Stuart DG (1976) Analysis of muscle receptor connections by spike-triggered averaging. 2. Spindle group II afferents. J Neurophysiol 39:1393–1402

Stephenson JBP (1978) Reversal of hypnosis-induced analgesia by naloxone. Lancet 2:991 – 992
Sternbach RA (1976) The need for an animal model of chronic pain. Pain 2:2 – 4
Sternbach RA, Ignelzi RJ, Deems LM, Timmerman G (1976) Transcutaneous electrical analgesia: a follow-up analysis. Pain 2:35 – 41
Stevens SS (1970) Neural events and the psychophysical law. Science 170:1043 – 1050
Stewart D, Thomson J, Oswald I (1977) Acupuncture analgesia: an experimental investigation. Br Med J 1:67 – 70
Strassburg HM, Krainick JU, Thoden U (1977) Influence of transcutaneous nerve stimulation (TNS) on acute pain. J Neurol 217:1 – 10
Sugrue MF (1974) The effects of acutely administered analgesics on the turnover of noradrenaline and dopamine in various regions of the rat brain. Br J Pharmacol 52:159 – 165
Swanson LW, McKellar P (1979) The distribution of oxytocin- and neurophysin stained fibers in the spinal cord of the rat and monkey. J Comp Neurol 188:87 – 106
Sweet WH (1977) Intracranial electrical stimulation for relief of chronic intractable pain. Prog Neurol Surg 8:258 – 269
Sweet WH, Wepsic JG (1968) Treatment of chronic pain by stimulation of fibers of primary afferent neuron. Trans AM Neurol Assoc 93:103 – 105
Sweet WH, Wepsic JG (1974) Stimulation of the posterior columns of the spinal cord for pain control: indications, technique, and results. Clin Neurosurg 21:278 – 310
Takagi H, Nakama M (1968) Studies on the mechanism of action of tetrabenazine as a morphine antagonist. II. A participation of catecholamine in the antagonism. Jpn J Pharmacol 18:54 – 58
Takagi H, Matsumura M, Yanai A, Ogiu K (1955) The effect of analgesics on the spinal reflex activity of the cat. Jpn J Pharmacol 4:76 – 187
Takagi H, Takashima T, Kimura K (1964) Antagonism of the analgetic effect of morphine in mice by tetrabenazine and reserpine. Arch Int Pharmacodyn Ther 149:484 – 492
Takagi H, Satoh M, Akaike A, Shibata T, Yajima H, Ogawa H (1978) Analgesia by enkephalins injected into the nucleus reticularis gigantocellularis of rat medulla oblongata. Eur J Pharmacol 49:113 – 116
Takagi H, Shiomi H, Ueda H, Amano H (1979) A novel analgesic dipeptide from bovine brain is a possible Met-enkephalin releaser. Nature 282:410 – 412
Takagi H, Satoh M, Akaike A, Shibata T, Kuraishi Y (1977) The nucleus reticularis gigantocellularis of the medulla oblongata is a highly sensitive site in the production of morphine analgesia in the rat. Eur J Pharmacol 45:91 – 92
Takemori AE (1974) Determination of pharmacological constants: use of narcotic antagonists to characterize analgesic receptors. Adv Biochem Psychopharmacol 8:335 – 344
Takemori AE, Tulunay FC, Yano I (1975) Differential effects on morphine analgesia and naloxone antagonism by biogenic amine modifiers. Life Sci 17:21 – 28
Tang AH, Schoenfeld MJ (1978) Comparison of subcutaneous spinal subarachnoid injections of morphine and naloxone on analgesic tests in the rat. Eur J Pharmacol 52:215 – 223
Tapper DN, Brown PB, Moraff H (1973) Functional organization of the cat's dorsal horn: connectivity of myelinated fiber systems of hairy skin. J Neurophysiol 36:817 – 826
Taub A (1976) Acupuncture "anesthesia": a critical review. Adv Pain Res Ther 1:743 – 748
Taub A (1964) Local, segmental and supraspinal interaction with a dorsolateral spinal cutaneous afferent system. Exp Neurol 10:357 – 374
Taub A, Campbell JN (1974) Percutaneous local electrical analgesia: peripheral mechanisms. Adv Neurol 4:727 – 732
Tenen SS (1968) Antagonism of the analgesic effect of morphine and other drugs by *p*-chlorophenylalanine, a serotonin depletor. Psychopharmacologia 12:278 – 285

Terenius L (1977) Opioid peptides and opiates differ in receptor selectivity. Psychoneuroendocrinology 2:53 – 58

Terenius L (1978) Endogenous peptides and analgesia. Annu Rev Pharmacol Toxicol 18:189 – 204

Terenius L (1973a) Stereospecific interaction between narcotic analgesics and a synaptic plasma membrane fraction of rat cerebral cortex. Acta Pharmacol Toxicol 32:317 – 320

Terenius L (1973b) Characteristics of the "receptor" for narcotic analgesics in synaptic plasma membrane fraction from rat brain. Acta Pharmacol Toxicol 33:377 – 384

Terenius L, Wahlström A (1975) Morphine-like ligand for opiate receptors in human CSF. Life Sci 16:1759 – 1764

Teschemacher H (1978) Endogenous ligands of opiate receptors (endorphins). In: Herz A (ed) Developments in opiate research. Dekker, New York, pp 67 – 151

Teschemacher H, Opheim KE, Cox BM, Goldstein A (1975) A peptide-like substance from pituitary that acts like morphine. 1. Isolation. Life Sci 16:1771 – 1776

Teschemacher H, Schubert P, Herz A (1973) Studies concerning the supraspinal site of action of morphine when inhibiting the hindleg flexion reflex. Neuropharmacol 12:123 – 131

Teuber HL (1960) Perception. In Field J (ed) Neurophysiology, vol III. American Physiol Soc, Washington, DC (Handbook of physiology, sect 1, pp 1595 – 1668 (close parenthesis)

Thexton AJ (1974) Jaw opening and jaw closing reflexes in the cat. Brain Res 66:425 – 433

Thorsteinsson G, Stonnington HH, Stillwell GK, Elveback LR (1978) The placebo effect of transcutanous electrical stimulation. Pain 5:31 – 41

Thorsteinsson G, Stonnington HH, Stillwell GK, Elveback LR (1977) Transcutaneous electrical stimulation: a double-blind trial of its efficacy for pain. Arch Phys Med Rehabil 58:8 – 13

Thrush DC (1973) Congenital insensitivity to pain. Brain 96:369 – 386

Toda K, Ichioka M (1978) Electroacupuncture: relations between forelimb afferent impulses and suppression of jaw-opening reflex in the rat. Exp Neurol 61:465 – 470

Toda K, Suda H, Ichioka M, Iriki A (1980) Local electrical stimulation: effective needling points for suppressing jaw opening reflex in rat. Pain 9:199 – 207

Tohyama M, Sakai K, Salvert D, Touret M, Jouvet M (1979a) Spinal projections from the lower brain stem in the cat as demonstrated by the horseradish peroxidase technique. I. Origins of the reticulospinal tracts and their funicular trajectories. Brain Res 173:383 – 403

Tohyama M, Sakai T, Touret M, Salvert D, Jouvet M (1979b) Spinal projections from the lower brain stem in the cat as demonstrated by the horseradish peroxidase technique. II. Projections from the dorsolateral pontine tegmentum and raphe nuclei. Brain Res 176:215 – 231

Torebjörk HE, Hallin RG (1974) Excitation failure in thin nerve fiber structures and accompanying hypalgesia during repetitive electric skin stimulation. Adv Neurol 4:733 – 735

Torvik A, Brodal A (1957) The origin of reticulospinal fibers in the cat. An experimental study. Anat Rec 128:113 – 138

Towe AL (1973) Somatosensory cortex: descending influences on ascending systems. In: Iggo A (ed) Somatosensory system. Springer, Berlin Heidelberg New York (Handbook of sensory physiology, vol 2, pp 701 – 718)

Towe AL, Jabbur SJ (1961) Cortical inhibition of neurons in dorsal column nuclei of cat. J Neurophysiol 24:488 – 498

Trevino DL, Coulter JD, Willis WD (1973) Location of cells of origin of spinothalamic tract in lumbar enlargement of the monkey. J Neurophysiol 36:750 – 761

Trevino DL, Maunz RA, Bryan RN, Willis WD (1972) Location of cells of origin of the spinothalamic tract in the lumbar enlargement of cat. Exp Neurol 34:64 – 77

Truex RC, Taylor MJ, Smythe MQ, Gildenberg P (1965) The lateral cervical nucleus of cat, dog and man. J Comp Neurol 139:93 – 104

Tseng LF, Loh HH, Li CH (1976) β-Endorphin as a potent analgesic by intravenous injection. Nature 263:239 – 240

Tsou K, Jang CS (1964) Studies on the site of analgesic action of morphine by intracerebral micro-injection. Sci Sin 13:1099 – 1109

Tulunay FC, Yano I, Takemori AE (1976) The effect of biogenic amine modifiers on morphine analgesia and its antagonism by naloxone. Eur J Pharmacol 35:285 – 292

Uddenberg N (1968) Functional organization of long, second-order afferents in the dorsal funiculus. Exp Brain Res 4:377 – 382

Uhl GR, Goodman RR, Kuhar MJ, Childers SR, Snyder SH (1979) Immunohistochemical mapping of enkephalin containing cell bodies, fibers and nerve terminals in the brain stem of the rat. Brain Res 166:75 – 94

Urca G, Frenk H, Liebeskind JC, Taylor AW (1978) Morphine and enkephalin: analgesic and epileptic properties. Science 197:83 – 86

Van der Ark GD, McGrath KA (1975) Transcutaneous electrical stimulation in treatment of postoperative pain. Am J Surg 130:338 – 340

Vedernikov YP (1969) Interaction of amphetamine, apomorphine, disulfiram with morphine and the role played by catecholamines in morphine analgesic action. Arch Int Pharmacodyn Ther 182:59 – 64

Vedernikov YP, Afrikanov II (1969) On the role of a central adrenergic mechanism in morphine analgesic action. J Pharm Pharmacol 21:845 – 847

Vierck CJ, Lineberry CG, Lee PK, Calderwood HW (1974) Prolonged hypalgesia following "acupuncture" in monkeys. Life Sci 15:1277 – 1289

Vierck CJ, Luck MM (1979) Loss and recovery of reactivity to noxious stimuli in monkeys with primary spinothalamic cordotomies, followed by secondary and tertiary lesions of other cord sectors. Brain 102:233 – 248

Vigouret J, Teschemacher H, Albus K, Herz A (1973) Differentiation between spinal and supraspinal sites of action of morphine when inhibiting the hindleg flexor reflex in rabbits. Neuropharmacol 12:111 – 121

Vogt M (1974) The effect of lowering the 5-hydroxytryptamine content of the rat spinal cord on analgesia produced by morphine. J Physiol (Lond) 236:483 – 498

Von Graffenried B, Del Pozo E, Roubicek J, Krebs E, Pöldinger W, Burmeister P, Kerp L (1978) Effects of the synthetic enkephalin analogue FK 33-824 in man. Nature 272:729 – 730

Von Knorring L, Alamay BGL, Johansson F, Terenius L (1978) Pain perception and endorphin levels in cerebrospinal fluid. Pain 5:359 – 366

Voorhoeve PE, Laporte Y, Bessou P (1958) A flexor reflex provoked by stimulation of non-medullated C-fibres in the cat. Acta Physiol Pharmacol Neerl 8:299 – 300

Wagman IH, Price DD (1969) Responses of dorsal horn cells of *M. mulatta* to cutaneous and sural nerve A and C fiber stimuli. J Neurophysiol 32:803 – 817

Wahlström A, Brandt M, Moroder L, Wünsch E, Lindeberg G, Ragnarsson U, Terenius L, Hamprecht B (1977) Peptides related to β-lipotropin with opioid activity. FEBS Lett 77:28 – 32

Walberg F (1957) Corticofugal fibres to the nuclei of the dorsal columns. An experimental study in the cat. Brain 80:273 – 287

Walberg F (1965) Axoaxonic contacts in the cuneate nucleus, probable basis for presynaptic depolarization. Exp Neurol 13:218 – 231

Walker JM, Berntson GG, Sandman CA, Coy DH, Schally AV, Kastin AJ (1977) An analog of enkephalin having prolonged opiate-like effects *in vivo*. Science 196:85 – 87

Walker RJ, Willis JC, Willis WD (1970) The action of para-methoxyphenylethylamine (PMPEA) on monosynaptic reflex transmission in the cat. Br J Pharmacol 38:106 – 116

Wall PD (1967) The laminar organization of dorsal horn and effects of descending impulses. J Physiol (Lond) 188:403 – 423
Wall PD (1975) Editorial. Pain 1:1 – 2
Wall PD (1976) Editorial. Pain 2:1
Wall PD, Merrill EG, Yaksh TL (1979) Responses of single units in laminae 2 and 3 of cat spinal cord. Brain Res 160:245 – 260
Wall PD, Sweet WH (1967) Temporary abolition of pain in man. Science 155:108 – 109
Wang JK (1977) Antinociceptive effect of intrathecally administered serotonin. Anesthesiology 47:269 – 271
Wang JK, Nauss LA, Thomas JE (1979) Pain relief by intrathecally applied morphine in man. Anesthesiology 50:149 – 151
Waterfield AA, Hughes J, Kosterlitz HW (1976) Cross tolerance between morphine and methionine-enkephalin. Nature 260:624 – 625
Watkins LR, Griffin G, Leichnetz GR, Mayer DJ (1980) The somatotopic organization of the nucleus raphe magnus and surrounding brain stem structures as revealed by HRP slow-release gels. Brain Res 181:1 – 16
Watson SJ, Akil H, Sullivan S, Barchas JD (1977) Immunocytochemical localization of methionine enkephalin: Preliminary observations. Life Sci 21:733 – 738
Watson SJ, Akil H, Richard CW, Barchas JD (1978) Evidence for two separate opiate peptide neuronal systems. Nature 275:226 – 228
Way EL (1972) Role of serotonin in morphine effects. Fed Proc 31:113 – 120
Weight FF, Salmoiraghi GC (1966) Response of spinal cord interneurons to acetylcholine, norepinephrine and serotonin administered by microelectrophoresis. J Pharmacol Exp Ther 153:420 – 427
Weil-Fugazza J, Godefroy F, Coudert D, Besson JM (1981) Morphine analgesia and newly synthesized 5-hydroxytryptamine in the dorsal and the ventral halves of the spinal cord of the rat. Brain Res 214:440 – 444
Weisberg JA, Rustioni A (1976) Cortical cells projecting to the dorsal column nuclei of cats. An anatomical study with the horseradish peroxidase technique. J Comp Neurol 168:425 – 438
Weisberg JA, Rustioni A (1977) Cortical cells projecting to the dorsal column nuclei of rhesus monkeys. Exp Brain Res 28:521 – 528
Weiss B, Laties VG (1958) Fractional escape and avoidance on a titration schedule. Science 128:1575 – 1576
Weiss B, Laties VG (1961) Changes in pain tolerance and other behavior produced by salicylates. J Pharmacol Exp Ther 131:120 – 129
Weiss B, Laties VG (1964) Analgesic effects in monkeys of morphine, nalorphine, and a benzomorphan narcotic antagonist. J Pharmacol Exp Ther 143:169 – 173
Weitzman ED, Ross GS (1962) A behavioral method for the study of pain perception in the monkey. Neurology 12:264 – 272
Werner G, Mountcastle VB (1965) Neural activity in mechanoreceptive cutaneous afferents: stimulus-response relations, Weber functions, and information transmission. J Neurophysiol 28:359 – 397
Wesche D, Höllt V, Herz A (1977) Radioimmunoassay of enkephalins. Regional distribution in rat brain after morphine treatment and hypophysectomy. Naunyn-Schmiedebergs Arch Pharmakol 301:79 – 89
Wessendorf MW, Proudfit HK, Anderson EG (1981) The identification of serotonergic neurons in the nucleus raphe magnus by conduction velocity. Brain Res 214:168 – 173
West DC, Wolstencroft JH (1977) Location and conduction velocity of raphe-spinal neurones in nucleus raphe magnus and raphe pallidus in the cat. Neurosci Lett 5:147 – 151

Westlund KN, Bowker RM, Ziegler MG, Coulter JD (1981) Origins of spinal noradrenergic pathways demonstrated by retrograde transport of antibody to dopamine-β-hydroxylase. Neurosci Lett 25:243–249

Westlund KN, Coulter JD (1980) Descending projections of the locus coeruleus and subcoeruleus/medial parabrachial nuclei in monkey: axonal transport studies and dopamine-β-hydroxylase immunocytochemistry. Brain Res Rev 2:235–264

White JC, Sweet WH (1955) Pain. Its mechanisms and neurosurgical control. CC Thomas, Springfield, Ill,

White JC, Sweet WH (1969) Pain and the neurosurgeon. CC Thomas, Springfield, Ill

Willer JC (1977) Comparative study of perceived pain and nociceptive flexion reflex in man. Pain 3:69–80

Willer JC, Boureau F, Albe-Fessard D (1979) Supraspinal influences on nociceptive flexion reflex and pain sensation in man. Brain Res 179:61–68

Willis WD (1981) Ascending pathways from the dorsal horn. In: Brown AG, Rethelyi M (eds): Spinal cord sensation. Scottish Academic Press, Edinburgh, pp 169–178

Willis WD (to be published 1983) The spinothalamic tract. In: Rosenberg R: The clinical neurosciences, sect V, Neurobiology. Churchill Livingstone, New York

Willis WD, Coggeshall RE (1978) Sensory mechanisms of the spinal cord. Plenum Press, New York

Willis WD, Haber LH, Martin RF (1977) Inhibition of spinothalamic tract cells and interneurons by brain stem stimulation in the monkey. J Neurophysiol 40:968–981

Willis WD, Kenshalo DR, Jr, Leonard RB (1979) The cells of origin of the primate spinothalamic tract. J Comp Neurol 188:543–574

Willis WD, Maunz RA, Foreman RD, Coulter JD (1975) Static and dynamic responses of spinothalamic tract neurons to mechanical stimuli. J Neurophysiol 38:587–600

Willis WD, Trevino DL, Coulter JD, Maunz RA (1974) Responses of primate spinothalamic tract neurons to natural stimulation of hindlimb. J Neurophysiol 37:358–372

Wilson CWM (1962) Suggestion and the placebo: an analysis of bias in clinical trials. In: Keele CA, Smith R (eds) The assessment of pain in man and animals. Livingstone, London, pp 213–228

Wilson VJ, Kato M (1965) Excitation of extensor motoneurons by group II afferent fibers in ipsilateral muscle nerves. J Neurophysiol 28:545–554

Winder CV, Pfeiffer CC, Maison GL (1946) The nociceptive contraction of the cutaneous muscle of the guinea pig as elicited by radiant heat. Arch Int Pharmacodyn Ther 72:329–359

Winter DL (1965) N gracilis of cat. Functional organization and corticofugal effects. J Neurophysiol 28:48–70

Wolf S (1959) The pharmacology of placebos. Pharmacol Rev 11:689–704

Wolf S, Hardy JD (1941) Studies on pain. Observations on pain due to local cooling and on factors involved in the "cold pressor" effect. J Clin Invest 20:521–533

Wolff BB (1971) Factor analysis of human pain responses: pain endurance as a specific pain factor. J Abnorm Psychol 78:292–298

Wolff BB (1978) Behavioral measures of human pain. In: Sternbach RA (ed) The psychology of pain. Raven Press, New York, pp 129–168

Wolff HG, Goodell H (1943) The relation of attitude and suggestion to the perception of and reaction to pain. Proc Assoc Res Nerv Ment Dis 23:434–446

Wolff HG, Hardy JD (1947) On the nature of pain. Physiol Rev 27:167–199

Wolff HG, Hardy JD, Goodell H (1940) Studies on pain. Measurement of the effect of morphine, codeine, and other opiates on the pain threshold and an analysis of their relation to the pain experience. J Clin Invest 19:659–680

Wolff HG, Wolf S (1948) Pain. CC Thomas, Springfield

Woodfore JM, Merskey H (1972) Some relationships between subjective measures of pain. J Psychosom Res 16:173–178

Woolf CJ (1979) Transcutaneous electrical nerve stimulation and the reaction to experimental pain in human subjects. Pain 7:115–128

Woolf CJ, Barrett GD, Mitchell D, Myers RA (1977) Naloxone-reversible peripheral electroanalgesia in intact and spinal rats. Eur J Pharmacol 45:311–314

Woolf CJ, Mitchell D, Barrett GD (1980) Antinociceptive effect of peripheral segmental electrical stimulation in the rat. Pain 8:237–252

Woolfe G, MacDonald AD (1944) The evaluation of the analgesic action of pethidine hydrochloride (Demerol). J Pharmacol Exp Ther 80:300–307

Yaksh TL (1978) Analgetic actions of intrathecal opiates in cat and primate. Brain Res 153:205–210

Yaksh TL (1979) Direct evidence that spinal serotonin and noradrenaline terminals mediate the spinal antinociceptive effects of morphine in the periaqueductal gray. Brain Res 160:180–185

Yaksh TL, Henry JL (1978) Antinociceptive effects of intrathecally administered human β-endorphin in the rat and cat. Can J Physiol Pharmacol 56:754–759

Yaksh TL, Rudy TA (1976) Analgesia mediated by a direct spinal action of narcotics. Science 192:1357–1358

Yaksh TL, Rudy TA (1977) Studies on the direct spinal action of narcotics in the production of analgesia in the rat. J Pharmacol Exp Ther 202:411–428

Yaksh TL, Rudy TA (1978) Narcotic analgetics: CNS sites and mechanisms of action as revealed by intracerebral injection techniques. Pain 4:299–359

Yaksh TL, Tyce GM (1979) Microinjection of morphine into the periaqueductal gray evokes the release of serotonin from the spinal cord. Brain Res 171:176–181

Yaksh TL, Wilson PR (1979) Spinal serotonin terminal system mediates antinociception. J Pharmacol Ther 208:446–453

Yaksh TL, DuChateau JC, Rudy TA (1976a) Antagonism by methysergide and cinanserin of the antinociceptive action of morphine administered into the periaqueductal gray. Brain Res 104:367–372

Yaksh TL, Yeung JC, Rudy TA (1976b) Systematic examination in the rat of brain sites sensitive to the direct application of morphine: observation of differential effects within the periaqueductal gray. Brain Res 114:83–103

Yaksh TL, Yeung JC, Rudy TA (1976c) An inability to antagonize with naloxone the elevated nociceptive thresholds resulting from electrical stimulation of the mesencephalic central gray. Life Sci 18:1193–1198

Yaksh TL, Plant RL, Rudy TA (1977a) Studies on the antagonism by raphe lesions of the antinociceptive action of systemic morphine. Eur J Pharmacol 41:399–408

Yaksh TL, Huang SP, Rudy TA, Frederickson RCA (1977b) The direct and specific opiate-like effect of met^5-enkephalin and analogues on the spinal cord. Neuroscience 2:593–596

Yang HY, Hong JS, Costa E (1977) Regional distribution of leu and met enkephalin in rat brain. Neuropharmacol 16:303–307

Yamashiro D, Tseng LF, Li CH (1977) [D-thr^2, thz^5]- and [D-met^2, thz^5]-enkephalinamides: Potent analgesics by intravenous injection. Biochem Biophys Res Commun 78:1124–1129

Yarbrough GG, Buxbaum DM, Sanders-Bush E (1972) Increased serotonin turnover in acutely morphine-treated mice. Biochem Pharmacol 21:2667–2669

Yarbrough GG, Buxbaum DM, Sander-Bush E (1973) Biogenic amines and narcotic effects. II. Serotonin turnover in the rat after acute and chronic morphine administration. J Pharmacol Exp Ther 185:328

Yeung JC, Yaksh TL, Rudy TA (1977) Concurrent mapping of brain sites for sensitivity to the direct application of morphine and focal electrical stimulation in the production of antinociception in the rat. Pain 4:23–40

Yezierski RP, Gerhart KD, Willis WD (1981) A further examination of the effects of cortical stimulation on primate spinothalamic tract cells (abstr). Soc Neurosci 7:228

Yezierski RP, Wilcox TK, Willis WD (1982) The effects of serotonin antagonists on the inhibition of primate spinothalamic tract cells produced by stimulation in nucleus raphe magnus or the periaqueductal gray. J Pharmacol Exp Ther 220:266 – 277

Yi CC, Lu TH, Wu SH, Tsou K (1977) A study of the release of ^{3}H-5-hydroxytryptamine from brain during acupuncture and morphine analgesia. Sci Sin 20:113 – 124

York JL, Maynert EW (1978) Alterations in morphine analgesia produced by chronic deficits of brain catcholamines or serotonin: role of analgesimetric procedure. Psychopharmacology 56:119 – 125

Yoss RE (1953) Studies of the spinal cord. Part 3. Pathways for deep pain within the spinal cord and brain. Neurology 3:163 – 175

Yu SKJ, Schmidt A, Sessle BJ (1973) Inhibitory effects on jaw muscle activity of innocuous and noxious stimulation of facial and intraoral sites in man. Arch Oral Biol 18:861 – 870

Yunger LM, Harvey JA (1976) Behavioral effects of L-5-hydroxytryptophan after destruction of ascending serotonergic pathways in the rat: the role of catecholaminergic neurons. J Pharmacol Exp Ther 196:307 – 315

Yunger LM, Harvey JA, Lorens SA (1973) Dissociation of the analgesic and rewarding effects of brain stimulation in the rat. Physiol Behav 10:909 – 913

Zhang A, Pan X, Xu S, Cheng J, Mo W (1980) Endorphins and acupuncture analgesia. Chin Med J 93:673 – 680

Zieglgänsberger W, Bayerl H (1976) The mechanism of inhibition of neuronal activity by opiates in the spinal cord of cat. Brain Res 115:111 – 128

Zieglgänsberger W, Tulloch IF (1979) The effects of methionine- and leucine-enkephalin on spinal neurones of the cat. Brain Res 167:53 – 64

Zorman G, Hentall ID, Adams JE, Fields HL (1981) Naloxone-reversible analgesia produced by microstimulation in the rat medulla. Brain Res 219:137 – 148

Subject Index

Handbook of Sensory Physiology

Springer-Verlag
Berlin
Heidelberg
New York

Volume 1
Principles of Receptor Physiology
Editor: W. R. Loewenstein
1971. 262 figures. XII, 600 pages
ISBN 3-540-05144-9

Volume 5 (In 3 Parts)
Auditory System
Part 1:
Anatomy, Physiology (Ear)
Editors: W. D. Keidel, W. D. Neff
1974. 305 figures. VIII, 736 pages
ISBN 3-540-06676-4

Part 2:
Physiology (CNS), Behavioral Studies, Psychoacoustics
Editors: W. D. Keidel, W. D. Neff
1975. 209 figures, 12 tables. VII, 526 pages
ISBN 3-540-07000-1

Part 3:
Clinical and Special Topics
Editors: W. D. Keidel, W. D. Neff
1976. 343 figures. VII, 811 pages
ISBN 3-540-07129-6

Volume 6:
Vestibular System
Part 1
Basic Mechanisms
Editor: H. H. Kornhuber
1974. 251 figures. VIII, 676 pages
ISBN 3-540-06889-9

Part 2:
Psychophysics, Applied Aspects and General Interpretations
Editor: H. H. Kornhuber
1974. 198 figures. VIII, 680 pages
ISBN 3-540-06864-3

Volume 9:
Development of Sensory Systems
Editor: M. Jacobson
1978. 231 figures, 9 tables. IX, 469 pages
ISBN 3-540-08632-3

MIX
Papier aus verantwortungsvollen Quellen
Paper from responsible sources
FSC® C105338

If you have any concerns about our products,
you can contact us on
ProductSafety@springernature.com

In case Publisher is established outside the EU,
the EU authorized representative is:
Springer Nature Customer Service Center GmbH
Europaplatz 3, 69115 Heidelberg, Germany

Printed by Libri Plureos GmbH
in Hamburg, Germany